Surface Waves and Discontinuities

Surface Waves and Discontinuities

Peter Malischewsky

With 86 Figures and 9 Tables

Akademie-Verlag Berlin 1987

Verfasser:

Dr. rer. nat. Peter Malischewsky

Akademie der Wissenschaften der DDR
Zentralinstitut für Physik der Erde
Institutsteil Jena

Akademie-Verlag Berlin ISBN 3-05-500277-6
Elsevier Amsterdam Oxford New York Tokyo ISBN 0-444-98959-5

Erschienen im Akademie-Verlag Berlin, DDR-1086 Berlin, Leipziger Straße 3—4

Lizenznummer: 202 · 100/449/87
Printed in the German Democratic Republic
Gesamtherstellung: VEB Druckhaus „Maxim Gorki", 7400 Altenburg
Lektor: Dipl.-Met. Heide Deutscher
LSV 1435
Bestellnummer: 763 555 6 (6932)

03500

Preface

It is not easy to write a monograph devoted on the one hand to a very specific area while being of use to a much broader circle of users on the other. I leave it to the reader to decide whether this aim has been achieved to a certain extent in the present treatise upon surface waves propagating in media with lateral inhomogeneities. The main interest of the author lies in the field of seismology. However, the field of surface waves is very suitable for a demonstration of the cross fertilization of ideas in different areas of application of wave theory. The monograph takes into account the rapid progress made in applying surface waves in many fields of physics and engineering science during the last two decades (e.g. acoustic surface wave filters, non-destructive testing, optical wave-guides, geophysics). That the fundamentals of surface waves have been a well-founded part of wave theory for many years is beyond dispute. On the other hand the application is connected with various problems which cannot be solved by using the conventional theory alone, such as input and output of surface wave energy in communication engineering devices, low-loss transfer of energy in wave-guides, the investigation of the Earth's interior using seismic surface waves, amongst others. One crucial problem common to these fields is the question of the interaction of surface waves with discontinuities. The primary objective of this book, dealt with in chapter 1, is to treat this very problem in detail by using a very simple model, the so-called impedance surface. Although a whole series of monographs exists in international literature devoted to the propagation of waves in disturbed media, there is no cross branch treatise for surface waves considering seismological aspects as well. I am attempting here for the first time to take into account most recent research results from all parts of the world. Additionally, numerous unpublished contributions of the author working in this field for many years have been included. A speciality of the book is that various calculations are presented in great detail so that the "know-how" is accessible for beginners in this area.

I hope that the book serves a dual purpose. In addition to being a monograph for seismologists, wave theorists and users of wave-guides in the broad sense, it is also intended to be a textbook for post-graduate students interested in special applications of mathematical physics to wave-guide problems. Furthermore, the interested reader will find many references from all fields of research covering a period of nearly 100 years.

I would like to thank the many colleagues, both in the GDR and abroad, who have helped in the success of the publication.

I dedicate this book to the memory of my father Sigismund Malischewsky.

PETER MALISCHEWSKY

To the reader's attention

For typographic reasons, in many cases the Greek mode index stands top left from the corresponding quantity. That means, it should not be confused with preceding factors, for example:

$$\mu^{\nu}U^{\mu}U^{*} \overset{!}{=} \mu({}^{\nu}U)\,({}^{\mu}U)^{*},$$

$$\cos 2^{\nu}\gamma H \overset{!}{=} \cos 2({}^{\nu}\gamma)\,H.$$

Contents

1. Introduction

1.1. General background

Since we have a picture of the Earth consisting of more or less concentric shells with different physical and chemical properties, the theory of propagation of elastic waves in inhomogeneous media has become an important subject for seismology. A special problem of this theory is the propagation of waves in wave-guides. The theory of open wave-guides is of exceptional importance to this. The reason for this is that for different seismological applications the Earth can be approximated by a layered, elastic half-space which is infinitely deep. Sources of natural and artificial earthquakes emit P- and S-waves forming surface waves at a definite distance away from the source. The surface waves represent the most significant contribution of the recorded ground displacement during an earthquake.

The Earth is a high-dispersive system. A temporally short impulse in the source region reaches the distant observer as a disturbance of a long duration. This is a consequence of dispersion which is a typical wave-guide phenomenon which particularly effects the surface waves. For this reason it is not surprising that surface waves are suited to treatment using the theory of wave-guides, and this will also be carried out in this book. The investigation of seismic surface waves was initiated by the classical work of Rayleigh (1887) who had found a solution of the elastodynamic equation of motion, which described a wave phenomenon concentrated on the free surface of an homogeneous, elastic half-space. Further investigations of guided waves in simple structures followed. Worth mentioning are Love's (1911) paper concerning guided SH-waves in a homogenous layer over a homogeneous half-space and Stoneley's (1924) concerning guided P-SV-waves at the boundary of two welded half-spaces. The first discussion of the propagation of surface waves in a more complicated multilayered medium was carried out by Sezawa (1927). This formed a basis for interpreting that part of the seismogram formed by surface waves. One of the simplest, but most important problems of wave propagation had been solved in the last century, namely the reflection and diffraction of homogeneous waves at a plane, infinite boundary. This was a basic requirement for considerable progress in interpreting records of body waves. A more difficult task is the investigation of surface wave seismograms. Regardless of this, the dispersion of surface waves and related phenomena are now understood quite well. However, as a result of an enhanced interest in clarifying the structure of the Earth's interior it has become necessary to investigate the surface wave seismograms in a manner, going beyond dispersion. One important way of doing this is by incorporating the influence of lateral inhomogeneities.

In geophysics, the theory of plate tectonics has drawn attention away from regions of the Earth which can be considered laterally homogeneous, and focused attention upon boundaries between plates, e.g., boundaries made of trenches, mid-continent rifts and mountain chains. These regions are clearly not laterally homogeneous. Generally speaking there are deviations from the ideal wave-guide and their influence on the propagation of surface waves has to be considered. Such deviations can be caused, for example by a mismatch of the wave-guide, caused in turn by a fault, by vertical or sloping discontinuities or also by an embedded disturbing body (see Fig. 1).

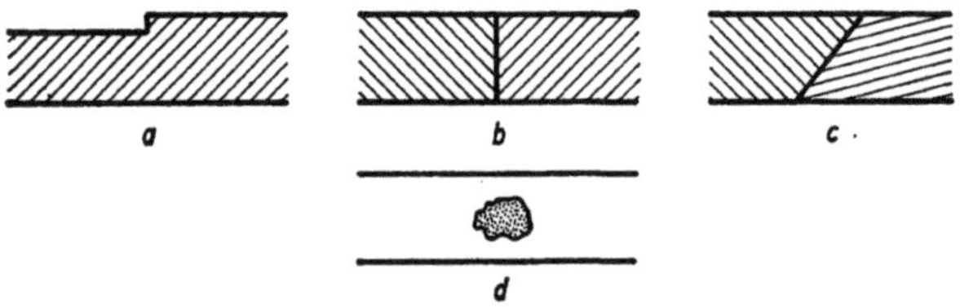

Fig. 1. Simple discontinuities within a wave-guide

a: step, b: vertical discontinuity, c: sloping discontinuity, d: embedded disturbing body

The treatment of the propagation of waves in wave-guides with disturbances requires very complicated mathematics. The publication of the author Malischewsky (1973) contains many references reflecting the level attained at that time. In the meantime, a great number of further theoretical as well as experimental papers have been published emphasizing the increasing interest in these problems. This interest comes from various fields of physics, geophysics, and technology.

The mathematical techniques developed for one field of application may also be useful for others. This is hardly surprising if one considers that waves, be they seismic, acoustic, electromagnetic, or others, have common mathematical features. The report of the Workshop "Application of Elastic Waves in Electrical Devices, Non-Destructive Testing, and Seismology" (1976) gives a good survey of the cross connections in applying the theory of elastic waves. A broad technical application of elastic surface waves was caused by their use in delay lines and transversal filter structures. As in seismology, the understanding of the propagation of surface waves in disturbed media has become important in this connection. The losses due to reflection and conversion into body waves, which occur as a result of the inhomogeneities, are of special interest.

As a result of the manifold technical applications the theory of electromagnetic wave-guides was elaborated very thoroughly. In recent years the ever-increasing application of optical wave-guides and integrated optics has resulted in rapid progress in this field, and in this connection there is a great interest in discontinuity problems, also from the theoretical and practical point of view. This can be seen if one considers the great number of publications devoted to these questions. Seismological literature has paid little attention to these papers. However, in the author's opinion it is just these papers which are particularly suited to developing the theory of surface waves for seismologically relevant questions. It cannot be the aim of this book to give a complete bibliography for elastic and electromagnetic wave-guides in general and discontinuities in particular. Those papers which are important for the development of ideas in this book have been quoted, as well as some further typical ones. An exceptional attention was paid to the monographies of MARCUSE (1972, 1974), MITTRA and LEE (1971), NEFEDOV (1979), SHEVCHENKO (1971a), and WEINSTEIN (1969). The interested reader can find many further references there. Although in the meantime solutions for rather complicated discontinuities have been obtained by using pure numerical methods on the basis of finite differences and finite elements [see e.g. BOORE (1970), LYSMER and DRAKE (1971), MUNASINGHE and FARNELL (1973), FUYUKI and MATSUMOTO (1980), MARTEL (1980), BOORE et al. (1981), SCHLUE (1979)], there is now as before, despite great efforts, a lack of practicable analytic or semi-analytic solutions. On the other hand, the search for analytical solutions is absolutely necessary to control the numerical solutions by using test cases. Otherwise the approximate solution is strictly speaking a numerical experiment which has been characterized by Samarsky by the words [see NEFEDOV (1979)]: "Without previous examination of the quality of the approximation procedures (but such an examination is possible only if the exact solution is known), it is very dangerous to use them because you cannot estimate whether the results of the numerical experiment reproduce the reality or any secondary effects of the approximation scheme."

For this reason, it is the main intention of this book to achieve a better understanding of the passage of seismic surface waves through discontinuities by using very simple analogy models which have, at least to some extent, analytical solutions. The models were selected by considering the principle: "As simple as possible, as complicated as necessary". In physics, again and again it has turned out to be a convenient way to begin the treatment of a specific problem by first of all considering a sequence of simpler problems with increasing complexity. Very often, an effect caused by a complex structure can be explained by superimposing effects of simpler structures. Of course, the solutions of the idealized models are primarily interesting as test cases, with the help of which the accuracy of the various approximation techniques can be estimated. But at the same time the rigorous solution is useful for clarifying the principal physical features of the phenomenon under consideration.

1.2. General remarks about wave-guides with discontinuities

Various methods have been developed for treating the propagation of guided waves (elastic surface waves, polaritons, channel waves, Lamb's waves and so on) in wave-guides with disturbances. To treat the problem in a mathematically rigorous way it would be necessary to solve a coupled linear system of partial differential equations with space-dependent coefficients. Disregarding the enormous mathematical difficulty, such a procedure would have the disadvantage that the physics of the interaction process between the waves and the disturbances would not become very apparent. Notwithstanding this fact, good results have been obtained in recent years, as already mentioned, by applying pure numerical methods. But in the course of this book our attention will be focused on analytical or semi-analytical methods. We distinguish here three groups on the whole

- methods which can be reduced to mode-matching techniques,
- methods which are based on optical ray theory,
- methods which use integral representation theorems.

Of course, there are various versions and combined methods. One characteristic of the first group is the introduction of a system of simple physical terms. Subsequently the expected processes are described within such a system. The basis of this system is formed by the modes of the undisturbed wave-guide, which are basic types of elastic or electromagnetic fields. They individually satisfy the boundary conditions. As is known, it is possible to represent the solutions of many problems of vibration theory by Fourier expansions in terms of the so-called fundamental oscillations of the system, which are eigenfunctions of any operator. If this operator looks like an ordinary linear differential equation of the second order with homogeneous boundary conditions at the ends of any interval, then we are dealing with a boundary value problem of the Sturm-Liouville type. The same procedure will be carried out for wave-guides.

Generally speaking, the mode-matching procedure can be applied advantageously if the geometry of the structure under consideration can be interpreted as a junction of two or more regions, at which each of them belongs to a separable coordinate system. The first step in applying the mode-matching procedure is to expand the unknown fields in the individual regions in terms of the respective normal modes. Proceeding on the assumption that the functional form of the normal modes is known, then the problem is reduced to the determination of the set of modal coefficients (or modal amplitudes), which is connected with the various regions. After that the continuity conditions have to be fulfilled on the planes of discontinuity and the orthogonality relation is used. This procedure leads to an infinite linear system of algebraic equations for the unknown modal expansion coefficients in the case of a closed wave-guide and to an integral equation in the case of an open wave-guide, respectively. The solution requires approximation techniques in each case. We will consider these questions in detail later on. According to YASUURA et al. (1980) the mode-matching technique is based on the Rayleigh principle for modal expansions of wave fields. This principle states that the infinite sum in modal expansions of fields can be used beyond its convergence area. For any field there exists an infinite sequence of truncated modal expansions converging uniformly in the entire domain under consideration. These authors confirmed that the mode-matching method is accurate and effective for numerical analysis, not only for closed wave-guides, but also for open ones. The mode conception has its limitations in two respects. Firstly, in the case of complicated boundary conditions it is no longer easily applicable because the modes themselves cannot be found. In these cases one can attempt to preserve the advantage of the mode conception by applying a method which is denominated coupled-mode theory in the English publications and cross-sectional method in the Soviet publications, respectively. Generally speaking, the fields in the complicated system are expanded in terms of a complete set of known modes of a simpler subsystem. These modes couple with one another, because they do not individually satisfy the boundary conditions of the complicated system.

Secondly, the modes describe the wave field globally, i.e., a given mode is connected with a functional dependence of the field defined over the whole cross-section of the wave-guide. If the field sought for differs considerably from each particular mode, many modes are required to represent the unknown field. A consequence of this is a loss of insight into the real local behaviour of the field, i.e. we have reached a limitation of the mode theory. Here it is advisable to use the methods of geometrical optics and WKBJ solutions, respectively.

It is well-known that there are two methods of interpreting a seismic signal, to give an example: the first one, as mentioned above, by superposition of normal modes, and the other one by a sum of reflected and refracted rays, which satisfy Snell's law. These two interpretations do not exclude but complement one another, because the spectrum of the normal modes is not practicable at high frequencies and the ray picture is only an approximation, which loses its sense at low frequencies.

The methods of the third group are based on completely different theoretical conceptions. On the one hand, all methods which try to solve the discontinuity problem by the Green's function technique belong in this group. On the other hand, there is the large group of methods based on the Fourier transform technique. According to MITTRA and LEE (1971) there are above all two methods of applying the Fourier transformation: the formulation as an integral equation, and Jone's formulation. In the final analysis, both variants lead to the application of the Wiener-Hopf technique. This technique makes it possible to avoid the analytical difficulties which are connected with the solution of certain types of boundary value problems using the integral transformations according to Fourier, Laplace, and Mellin, or with the transformation into integral equations. These problems belong mainly to the theory of wave-guides and to the diffraction of waves. As regards the Wiener-Hopf technique we only mention at this point, that the transformation variable in the integral transformation is assumed as complex and the principle of analytical continuation and Liouville's theorem are used. With the help of the Wiener-Hopf technique it is possible to obtain analytical solutions for a limited number of simple models. Among them there is the impedance jump on a surface. This simple model will play an important role as a crucial problem for the further understanding in what follows. This is why we will demonstrate the application of the Wiener-Hopf technique to this configuration in chapter 5 in great detail: the first time this has been attempted in seismological literature.

2. Mathematical aspects

2.1. The boundary conditions and their seismological interpretation

2.1.1. General remarks

In the following, all phenomena are described in a cartesian co-ordinate system, in which all quantities are independent of x_2, because we will focus our attention on the two-dimensional problem with a view to a considerable simplification. Let the waves propagate into the direction x_1 while x_3 is the transversal co-ordinate of the wave-guide. For all field quantities a time-dependence of the type $\exp(-i\omega t)$ is assumed, ω being the angular frequency, but the time factor is omitted in most cases for the sake of brevity. Further on it is assumed that Latin indices take the values 1, 2, 3 while Greek ones serve to number the eigenfunctions or modes. For Latin indices, Einstein's summation convention holds. An apostrophe indicates the derivative by the corresponding argument, if there is no other agreement.

In the course of this book all calculations are carried out for elastic waves. The same could be done in most cases with only slight modifications for many other kinds of waves. The plane $x_3 = 0$ is assumed to be the free surface of an ideally elastic half-space or the upper boundary of a wave-guide with finite diameter. While $u_i = u_i(x_1, x_3)$ $(i = 1, 2, 3)$ are the components of the displacement vector, let $U_i = U_i(x_3)$ be the x_3-dependences belonging to them, the so-called amplitude-depth-dependences and the eigenfunctions, respectively. Proceeding on the statement for plane waves $u_j(x_1, x_3; t) = U_j(x_3) \times \exp[i(kx_1 - \omega t)]$ the following boundary conditions have to be satisfied for $x_3 = 0$:

$$\begin{aligned} &\mu(0)\,[U_1'(0) + ikU_3(0)] = 0, \\ &\mu(0)\,U_2'(0) = 0, \qquad\qquad (2.1) \\ &[2\mu(0) + \lambda(0)]\,U_3'(0) + ik\lambda(0)\,U_1(0) = 0, \end{aligned}$$

expressing a stress-free surface. Here k is the wave number and $\mu = \mu(x_3)$, $\lambda = \lambda(x_3)$ are Lamé's elastic parameters. To begin with the boundary condition for U_2 alone, we see that it is not coupled with U_1 and U_3 and it holds for all SH-kind waves within the wave-guide. Surface waves of SH-type, i.e. Love waves, exist only then, if not $\beta'(x_3) \equiv 0$ for $x_3 \in [0, H]$, where $\beta = \beta(x_3)$ is the depth-dependent shear wave velocity and H is the thickness of the wave-guide. For the important special case of an open wave-guide, $H = \infty$ holds.

Love waves are very suitable for the investigation of the phenomena we are interested in, because only one component of the displacement vector is different from zero. This simple mathematical structure, however, is achieved at the expense of more complicated depth-dependence of the elastic parameters (in comparison to Rayleigh waves). It is well-known that the mathematically and physically simplest model admitting Love waves is "layer over half-space". In attempting to take advantage of the simple structure of Love waves and at the same time to maintain a simple depth-dependence of the elastic parameters we are led to a modification of the boundary conditions. Replacing the boundary condition $\mu(0)\, U_2'(0) = 0$ by

$$\mu(0)\, U_2'(0) + \varepsilon U_2(0) = 0 \quad \text{for} \quad x_3 = 0, \tag{2.2}$$

surface waves with $U_2 \neq 0$ also exist for elastic parameters which are constant over the whole cross-section of the wave-guide, as will be seen later. The quantity ε is a not necessary small positive parameter characterizing the surface impedance with $Im\varepsilon = 0$ and its dimension is stress/length. Equation (2.2) is an impedance-like boundary condition of the third kind, being equivalent to the boundary condition of Leontovich in the theory of electromagnetic waves. According to NEFEDOV (1979) this is given by

$$u + w \frac{\partial u}{\partial N} = 0 \quad \text{on} \quad S, \tag{2.3}$$

where N is the normal on the surface S and w is the surface impedance. For $|w| \to 0$ and $|w| \to \infty$, respectively, the boundary conditions of the first kind $u|_S = 0$ and of the second kind $\left.\frac{\partial u}{\partial N}\right|_S = 0$ are obtained. The surface impedance w may also have a more complicated structure, for example, in the case of anisotropy it can be a tensor. Using boundary conditions of the type (2.3), a great number of practical problems of diffraction theory can be solved. These boundary conditions are always important in those cases if one wants to describe the influence of a body on the wave field in its interior by an integral effect on its surface (so-called equivalent boundary conditions). On this basis, it is possible to find a subsequent seismological interpretation for the boundary condition (2.2), which was introduced in the first instance for purely mathematical reasons.

2.1.2. *Approximations of an elastic layer over an elastic half-space according to Tiersten (1969) and Alsop et al. (1971)*

It is well-known, that the investigation of surface wave propagation in layered elastic half-spaces is mathematically troublesome. In the case of a single layer over the half-space, the physically relevant information is already concealed within a determinant of the sixth order. However, if you are only interested in the influence of a very thin layer, in comparison with the wave length on surface wave propagation, the layer can be approximated according to Tiersten (1969) by special boundary conditions on the surface of the half-space. This is accomplished by using the approximate equations of low frequency extension and flexure of thin plates. The assumed geometry is shown in Fig. 2.

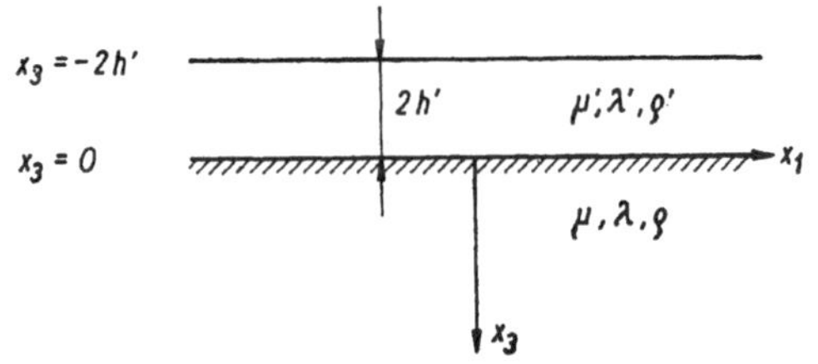

Fig. 2. Illustration of a thin layer over a half-space

The boundary conditions

$$\sigma_{3j} = 0 \quad \text{for} \quad x_3 = 0 \tag{2.4}$$

on the free surface are replaced by

$$\sigma_{3j} = -\delta_{jb}\, 2h'\mu' \left[\frac{2\mu' + 3\lambda'}{2\mu' + \lambda'}\, u_{a,a,b} + u_{b,a,a}\right] + 2h'\varrho'\ddot{u}_j$$
$$\text{for} \quad x_3 = 0, \qquad a, b = 1, 2; \qquad j = 1, 2, 3, \tag{2.5}$$

where σ_{ij} are the components of the stress tensor, μ, λ, ϱ are Lamé's parameters and the mass density of the half-space, respectively, μ', λ', ϱ' are the analogous quantities for the layer, and $2h'$ is its thickness. The apostrophes should not be confused here with the derivative. Substituting the corresponding expressions for Rayleigh and Love waves with u_j, Tiersten's boundary conditions may be written as follows (the summation convention does not hold for underlined indices)

$$\sigma_{i3} + \varepsilon_{\underline{i}} u_{\underline{i}} = 0 \quad \text{for} \quad x_3 = 0,$$
$$\varepsilon_1 = 2h'\varrho'\omega^2 \left[1 - \frac{4(\mu' + \lambda')}{(2\mu' + \lambda')} \frac{\beta'^2}{c_R^2}\right],$$
$$\varepsilon_2 = 2h'\varrho'\omega^2 \left[1 - \frac{\beta'^2}{c_L^2}\right] \tag{2.6}$$
$$\varepsilon_3 = 2h'\varrho'\omega^2.$$

Here c_R, c_L are the phase velocities of Rayleigh and Love waves respectively, and β' is the shear wave velocity of the layer. It can be seen that the quantity formerly introduced, ε may be immediately considered as being identical to ε_2. From Tiersten's boundary conditions the dispersion relation of Rayleigh waves is obtained as follows,

$$\omega^2 F(\xi) - \frac{\omega\beta}{\mu}\left(\varepsilon_3 \sqrt{1 - \frac{\beta^2}{\alpha^2}\xi^2} + \varepsilon_1 \sqrt{1 - \xi^2}\right)\xi^3$$

$$+ \frac{\varepsilon_1\varepsilon_3}{\mu^2}\beta^2\xi^2\left(1 - \sqrt{1 - \frac{\beta^2}{\alpha^2}\xi^2}\sqrt{1 - \xi^2}\right) = 0, \quad (2.7)$$

$$\xi = \frac{c_R}{\beta},$$

where α and β are the longitudinal and shear wave velocities of the half-space respectively, and $F(\xi)$ is the Rayleigh function defined by

$$F(\xi) = 4\sqrt{1 - \frac{\beta^2}{\alpha^2}\xi^2}\sqrt{1 - \xi^2} - (2 - \xi^2)^2. \quad (2.8)$$

Obviously, for $\varepsilon_1 = \varepsilon_3 \equiv 0$ equation (2.7) changes into the well-known dispersion relation

$$F(\xi) = 0 \quad (2.9)$$

of the half-space without layer. By using the abbreviations

$$r_\alpha = \sqrt{1 - \frac{c_R^2}{\alpha^2}}, \quad r_\beta = \sqrt{1 - \frac{c_R^2}{\beta^2}},$$

$$\gamma = \frac{4\pi h'}{T \cdot c_R} = \frac{2\omega h'}{c_R}, \quad A_0 = -4r_\alpha r_\beta + \left(2 - \frac{c_R^2}{\beta^2}\right)^2,$$

$$A_1 = \frac{\varrho' c_R^4}{\varrho\beta^4}\left[r_\alpha + \left(1 - \frac{B}{c_R^2}\right)r_\beta\right], \quad (2.10)$$

$$A_2 = -\frac{\varrho'^2 c_R^4}{\varrho^2\beta^4}\left(1 - \frac{B}{c_R^2}\right)(1 - r_\alpha r_\beta), \quad B = \frac{4\mu'(\mu' + \lambda')}{(2\mu' + \lambda')\varrho'}$$

it is possible to write equation (2.7) as

$$A_0 + A_1\gamma + A_2\gamma^2 = 0, \quad (2.11)$$

from which we conclude that for one period T there are two phase velocities c_R which are solutions of this quadratic equation. It is well-known that the model "layer over half-space" "produces" an infinite number of Rayleigh modes. Tiersten's approximation includes, in addition to the fundamental mode, the first higher mode at the very least, the so-called Sezawa mode [see SEZAWA and KANAI (1935)]. However, for all material parameters the latter is not a solution of (2.11). An elementary consideration shows, that for

$$\frac{\beta}{\beta'} > \sqrt{\frac{2}{1-\sigma'}}, \quad \sigma' = \frac{\lambda'}{2(\mu'+\lambda')} \tag{2.12}$$

only, two physically significant solutions of (2.11) exist, where σ' is Poisson's ratio of the layer. TIERSTEN (1969) did not refer to this fact.

For Love waves, Tiersten's boundary conditions yield the following dispersion relation

$$c_L^4\left(\frac{\omega^2}{\beta'^4}+\frac{A}{\beta^2}\right) - c_L^2\left(2\frac{\omega^2}{\beta'^2}+A\right) + \omega^2 = 0, \quad A = \frac{\varrho^2\beta^4}{\varrho'^2\beta'^4\, 4h'^2}, \tag{2.13}$$

which leads for $c_L > \beta'$ to only one significant solution as regards physics, corresponding to the fundamental Love mode. In Fig. 3 the approximated dispersion curves of Love waves for two models with different layer thicknesses are compared to the exact dispersion curves.

Obviously, with decreasing layer thickness and increasing period, Tiersten's boundary conditions describe the dispersion behaviour of Love waves for the model "layer over half-space" quite well.

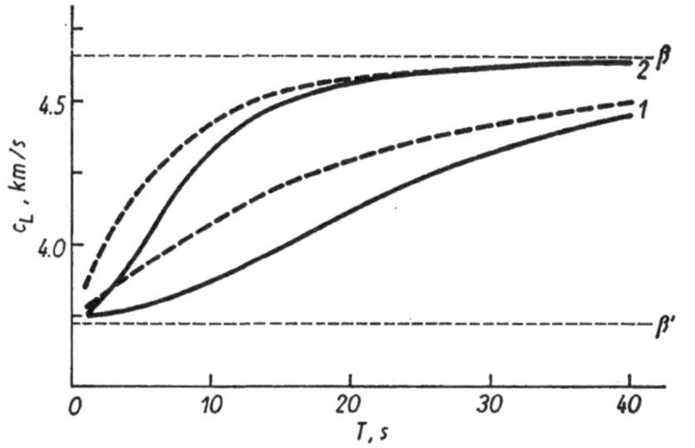

Fig. 3. Comparison of approximated and exact dispersion curves of Love waves (full lines: exact, dashed lines: Tiersten)

Model 1: $\beta' = 3.723$ km/s
$\varrho' = 2.77$ g/cm³
$h' = 15.0$ km
$\beta = 4.656$ km/s
$\varrho = 3.3$ g/cm³

Model 2: as model 1 but
$h' = 5.0$ km

For the description of the effect of a layer on surface wave propagation, a simpler approximation has been suggested by Alsop et al. (1971). These authors proceed from the assumption that this effect can be separated into terms arising from elastic and from inertial restraining forces. In a number of situations of practical importance, it turns out that the inertial terms are dominant. In these cases the concept of the so-called mass-loaded half-space is applicable. For Love-kind motion the boundary condition is obtained as

$$\sigma_{23} + 2h'\varrho'\omega^2 u_2 = 0, \qquad x_3 = 0,$$

and the dispersion relation is

$$c_L = \frac{\beta}{\sqrt{1 + \dfrac{h'^2\omega^2\varrho'^2}{\varrho^2\beta^2}}}. \tag{2.15}$$

According to Alsop et al. (1971) the mass loading approximation is excellent for describing the low-frequency dispersion of surface waves in a layer of high density and low rigidity with respect to the underlying substrat; that is if

$$\frac{\mu'}{\mu} \cdot \frac{\varrho}{\varrho'} \ll 1 \tag{2.16}$$

holds. Of course, Tiersten's approximation yields better results in each case.

To sum up, it can be said that the boundary conditions of type (2.3) have a quite practical significance, quite apart from their mathematical suitability.

2.2. The differential operators in closed domains

In the following chapters and in the course of this book we consider SH-like motion only, this being the simplest special case. The essentially more complicated SV-like motion is mentioned in those chapters where practical results are presented, and in chapter 6. The component u_2 of the displacement vector satisfies the partial differential equation

$$\frac{\partial}{\partial x_1}\left[\mu(x_3)\,\frac{\partial u_2}{\partial x_1}\right] + \frac{\partial}{\partial x_3}\left[\mu(x_3)\,\frac{\partial u_2}{\partial x_3}\right] = \varrho(x_3)\,\ddot{u}_2. \tag{2.17}$$

Let the functions $\mu(x_3)$, $\mu'(x_3)$, and $\varrho(x_3)$ be piecewise continuous (layered medium) and bounded in the considered domain of definition $0 \leqq x_3 \leqq H$. Moreover let $\mu(x_3)$ and $\varrho(x_3)$ be positive. By using the statement of plane waves

$$u_2(x_1, x_3, t) = U_2(x_3)\, e^{i(kx_1 - \omega t)}, \tag{2.18}$$

mentioned already in 2.1.1., the separation of variables leads to the ordinary differential equation

$$L_{22}U_2 = \mu k^2 U_2, \tag{2.19}$$

where L_{22} is the Love wave-operator defined by

$$L_{22} = \frac{\mathrm{d}}{\mathrm{d}x_3}\left(\mu \frac{\mathrm{d}}{\mathrm{d}x_3}\right) + \varrho\omega^2, \tag{2.20}$$

[compare with Eq. (8) in MALISCHEWSKY (1976)]. In the following, let $U_2 = U$ and $u_2 = u$ for the sake of brevity. Using the abbrevations

$$L = \frac{1}{\mu}\left[\frac{\mathrm{d}}{\mathrm{d}x_3}\left(\mu \frac{\mathrm{d}}{\mathrm{d}x_3}\right) + \varrho\omega^2\right], \qquad \lambda = k^2, \tag{2.21}$$

the differential equation (2.19) may be written as

$$LU = \lambda U. \tag{2.22}$$

The definition of an operator L has to be complemented by the modification of its domain of definition $\vartheta(L)$: that is the totality of all functions U, for which LU is defined. More generally as in (2.2) the boundary conditions are assumed to be

$$\begin{aligned} \nu_0 U'(0) + \nu_1 U(0) &= 0, \\ \eta_0 U'(H) + \eta_1 U(H) &= 0, \end{aligned} \tag{2.23}$$

where ν_0, ν_1, η_0, η_1 are real and neither $\nu_0 = \nu_1 = 0$ nor $\eta_0 = \eta_1 = 0$. At the layer boundaries, i.e. at the planes of discontinuity of the elastic parameters defined by $x_3 = z_k$, $k = 1, \ldots, N$ (see Fig. 4), let the following continuity relations

$$\begin{aligned} &{}^{(k)}U(z_k) = {}^{(k-1)}U(z_k), \\ &{}^{(k)}\mu(z_k)\, {}^{(k)}U'(z_k) = {}^{(k-1)}\mu(z_k)\, {}^{(k-1)}U'(z_k), \\ &k = 1, \ldots, N, \end{aligned} \tag{2.24}$$

hold.

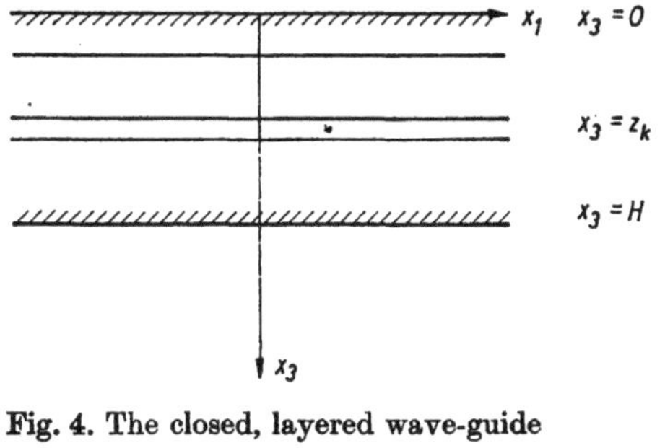

Fig. 4. The closed, layered wave-guide

The separable Hilbert space of complex measurable functions, for which $\int_0^H |f(z)|^2 \mu(z)\, \mathrm{d}z$ exists in the sense of Lebesgue, is denoted by $\mathbf{L}_2(0, H; \mu)$. Let (f, g) be the scalar product, i.e.,

$$\begin{aligned} \mathbf{L}_2(0, H; \mu) &= \left\{ f(z) \mid \int_0^H |f(z)|^2 \mu(z)\, \mathrm{d}z < \infty \right\}, \\ (f, g) &= \int_0^H f^*(z)\, g(z)\, \mu(z)\, \mathrm{d}z, \qquad f, g \in \mathbf{L}_2, \end{aligned} \tag{2.25}$$

where $\{f(z) \mid x\}$ should be read as the totality of all $f(z)$ with the attribute x. The operator L is now defined as

$$LU = \frac{1}{\mu} \left[\frac{\mathrm{d}}{\mathrm{d}x_3} \left(\mu \frac{\mathrm{d}}{\mathrm{d}x_3} \right) + \varrho\omega^2 \right] U \, , \quad \text{for} \quad U \in \vartheta(L), \tag{2.26}$$

with $\vartheta(L)$: 1) U, U' continuous and U'' piecewise continuous for $x_3 \in [0, H]$ & $x_3 \neq z_k$, $k = 1, \ldots, N$.
2) Let U satisfy the boundary conditions (2.23) and the continuity conditions (2.24).

Obviously, $\vartheta(L)$ is a subspace of $\mathbf{L}_2$, i.e. $\vartheta(L) \subseteq \mathbf{L}_2$. Under these assumptions, (2.22) is a regular, self-adjoint Sturm-Liouville eigenvalue problem with N points of discontinuity and corresponding interface conditions. According to Kazi (1976) the usual results of ordinary Sturm-Liouville theory for regular systems hold for discontinuous, regular, Sturm-Liouville systems with interface conditions after necessary modification. In particularly, the operator L has a discrete, real spectrum of eigenvalues λ_α ($\alpha = 1, 2, 3, \ldots$) with a single limit point at $\lambda = -\infty$. These eigenvalues form a numerable set, where, according to Courant and Hilbert (1968) and in keeping with Zavadsky (1972), only a finite number of positive eigenvalues $\lambda_1, \ldots, \lambda_\nu$ exists with a highest eigenvalue λ_1. The essentially negative character of the point spectrum also follows from the property of the considered operator L, of being semi-bounded from above. [Compare with Neumark (1960) by considering that if an operator A is semi-bounded from below, $-A$ is semi-bounded from above.] Consequently, the eigenvalues of L can be arranged as follows (see also Fig. 5):

$$\lambda_1 > \lambda_2 > \lambda_3 \cdots > \lambda_\nu > \lambda_{\nu+1} > \cdots. \tag{2.27}$$

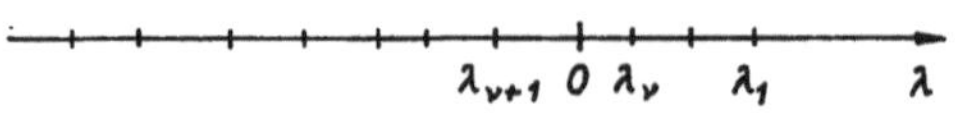

Fig. 5. The distribution of the discrete eigenvalues

Further, the operator L has a complete (total) system of eigenfunctions ${}^{\alpha}U$, satisfying the differential equation

$$L^{\alpha}U = \lambda_{\alpha}{}^{\alpha}U \tag{2.28}$$

as well as the orthonormality relation

$$({}^{\alpha}U, {}^{\beta}U) = \delta_{\alpha\beta}. \tag{2.29}$$

Each function $f, f \in \vartheta(L)$ can be expanded into uniform convergent generalized Fourier series in terms of these eigenfunctions of the operator L. For large values of γ the following asymptotic formulas for the eigenvalues and eigenfunctions hold [see SMIRNOV (1975)]:

$$\lambda_{\gamma} = -\frac{\pi^2\gamma^2}{H^2} + O(1), \quad {}^{\gamma}U = c \cdot \sin \frac{\mathrm{i}\pi x_3}{H} + O\left(\frac{1}{\gamma}\right). \tag{2.30}$$

Here, as usual, $O(1)$ is a quantity whose absolute value is bounded for all γ, and $O\left(\frac{1}{\gamma}\right)$ is a quantity for which the product $\gamma \cdot O\left(\frac{1}{\gamma}\right)$ is bounded for infinitely increasing γ.

2.3. The differential operators in open domains

The considerations of the preceding section are now to be extended to the limit case $H \to \infty$ (layered elastic half-space) we were originally interested in. Let us proceed from the differential operator L,

$$L = \frac{1}{\mu}\left[\frac{\mathrm{d}}{\mathrm{d}x_3}\left(\mu \frac{\mathrm{d}}{\mathrm{d}x_3}\right) + \varrho\omega^2\right], \quad x_3 \in [0, \infty), \tag{2.31}$$

being defined now in the Hilbert space $\mathbf{L}_2(0, \infty; \mu)$ over the semi-open interval $[0, \infty)$ with the same continuity properties of $\mu(x_3)$ and $\varrho(x_3)$ as previously. To specify its domain of definition, some additional considerations are necessary. The eigenvalue problem

$$LU = \lambda U, \qquad 0 \leqq x_3 < \infty \tag{2.32}$$

is a singular Sturm-Liouville problem, because the right endpoint is ∞. According to Weyl [see JÖRGENS and RELLICH (1976)] a singular differential operator of the second order can be in the limit-point or in the limit-circle case at the end-points of the interval (Weyl's alternative). These terms refer to the asymptotic behaviour of solutions of the eigenvalue problem in infinity, but a more detailed explanation cannot be given here. The operator of the regular Sturm-Liouville problem is in the limit-circle case at both end-points of the interval. The limit-circle case is also present at the left end-point $x_3 = 0$ of the interval under consideration. According to a limit point-limit circle criterion given by JÖRGENS and RELLICH (1976), the limit-point case is present at the right end-point $x_3 = \infty$, if

$$\frac{-\varrho\omega^2}{\mu} \geqq G > -\infty \quad \text{and with} \quad c \in (0, \infty),$$
$$g(z) = \int_c^z \frac{ds}{\mu(s)} \quad \text{it follows that} \quad g \notin \mathbf{L}_2(0, \infty; \mu) \tag{2.33}$$

holds. Assuming that there is a maximal shear wave velocity $\beta_{\max}$ and a maximal μ in the interval under consideration, we can set $G = -\omega^2/\beta_{\max}$ and due to $\lim\limits_{z\to\infty} g(z) = \infty$ it follows that $\int_c^\infty \mu g^2 \, dz = \infty$. That means, the limit-point case is present at $x_3 = \infty$ and the deficiency index of the operator L is equal to (1,1) in accordance with the theorem of Levinson [see NEUMARK (1960)]. To obtain a self-adjoint operator, $\vartheta(L)$ has possibly to be restricted by boundary conditions. It appears that only in the limit-circle case additional boundary conditions are necessary. Without discussing these things in detail, we state here that the first boundary condition of (2.23) at $x_3 = 0$ is exactly that one which allows a selfadjoint extension of the operator L. Thus, the domain of definition of the operator L is defined by

$\vartheta(L)$: 1) U, U' continuous, U'' piecewise continuous for $x_3 \in [0, \infty)$.
2) $U \in \mathbf{L}_2(0, \infty; \mu)$, $\quad LU \in \mathbf{L}_2(0, \infty; \mu)$. (2.34)
3) Let U satisfy the boundary condition $\nu_0 U(0) + \nu_1 U'(0) = 0$ and the continuity conditions (2.24).

The peculiarity of this singular Sturm-Liouville eigenvalue problem with N discontinuity points and continuity conditions lies in the fact that, contrary to the previous considerations, the operator L has a discrete spectrum (point spectrum) as well as a continuous one. The continuous spectrum occurs as a result of the infinite domain. We are far from being able to make such precise statements about the singular case of the infinite domain as we can about the regular one. Only sparse, or very general details concerning the case of mixed spectrum we are especially interested in have been published. Incidently, we cannot make general statements about the existence of discrete eigenvalues. In the following, let us assume that there is a finite number of discrete real eigenvalues λ_α, $\alpha = 1, \ldots, M$ for the configuration under consideration $\mu = \mu(x_3)$, $\varrho = \varrho(x_3)$ and for a given ω. Let us assume for the moment that $\varrho(x_3)$ vanishes near infinity, i.e. $\varrho(x_3) \equiv 0$ for $x_3 \in (\bar{c}, \infty)$ with $\bar{c} > 0$, and further $\mu(x_3)$ is constant in $(\bar{c}, \infty)$. Then, according to WEIDMANN's (1967) theorem 5.1 we are able to conclude that the spectrum of L is absolutely continuous in $(-\infty, 0)$, especially if there are no eigenvalues. According to WEIDMANN (1982) we can obtain further statements by applying the "decomposition method" [compare also with NEUMARK (1960)]. Let L_0 be the operator generated by the differential expression (2.31) in $\mathbf{L}_2(0, \bar{c}; \mu)$ with the boundary condition 3) at 0 and $U(\bar{c}) = 0$; let L_∞ be the operator generated in

Fig. 6. The spectrum of L in the open domain

$\mathbf{L}_2(\bar{c}, \infty; \mu)$ with the boundary condition $U(\bar{c}) = 0$. L_0 is regular, i.e. it has at best a finite number of positive eigenvalues. The operator L_∞ is a multiple of $\frac{\mathrm{d}^2}{\mathrm{d}x_3^2}$ in $\mathbf{L}_2(\bar{c}, \infty; \mu)$ with Dirichlet's condition at $\bar{c}$; thus, the spectrum of L_∞ is equal to $(-\infty, \bar{c}]$, i.e. it is equal to the continuous (essential) spectrum. Thus, in $\mathbf{L}_2(0, \infty; \mu) = \mathbf{L}_2(0, \bar{c}; \mu) \oplus \mathbf{L}_2(\bar{c}, \infty; \mu)$ the sum operator $L_0 \oplus L_\infty$ has the continuous spectrum $(-\infty, 0]$ and only a finite number of spectral points (= eigenvalues). L and $L_0 \oplus L_\infty$ are self-adjoint continuations of the operator L_{00} in $\mathbf{L}_2(0, \infty; \mu)$, which is defined by the boundary condition 3) at $x_3 = 0$ and $U(\bar{c}) = U'(\bar{c}) = 0$. According to WEIDMANN (1976) § 8.3, corollary 1, it follows that L has only a finite number of positive eigenvalues, too, while $(-\infty, 0]$ belongs entirely to the spectrum. Let us drop now the additional requirement about the vanishing of ϱ nearby ∞. Let $\lim\limits_{x_3 \to \infty} \varrho(x_3) = \varrho_\infty > 0$ be and $\lim\limits_{x_3 \to \infty} \mu(x_3) = \mu_\infty > 0$. Then following WEIDMANN (1967) the point $\Lambda = \omega^2/\beta_\infty^2$ with $\beta_\infty^2 = \mu_\infty/\varrho_\infty$ adopts the position of the remarkable point 0 within the spectrum, the spectrum of L is absolutely continuous in $(-\infty, \Lambda]$ and above Λ there is only a finite number of (discrete) eigenvalues λ_α, $\alpha = 1, \ldots, M$. Λ itself is never an eigenvalue. Fig. 6 shows the spectrum of L analogously to Fig. 5.

These statements also hold especially for the case where $\mu(x_3)$ and $\varrho(x_3)$ are piecewise constant within the considered domain with corresponding continuity conditions at the finite number of discontinuity points. Then $\Lambda = \omega^2/\beta_H{}^2$ holds, where β_H is the shear wave velocity of the half-space. This statement agrees with the papers of KAZI (1976) and KAZI and ABU-SAFIYA (1981).

where, as far as the spectrum of L is concerned, the same results for the special case of two layer- and three layer-models were obtained in a different way.

The presence of a mixed spectrum is connected with the existence of proper as well as improper eigenfunctions of the operator L. The proper eigenfunctions ${}^{\alpha}U$ belong to the discrete eigenvalues λ_α, and they satisfy the orthonormality relation

$$({}^{\alpha}U, {}^{\beta}U) = \int_0^\infty \mu\, {}^{\alpha}U^* {}^{\beta}U \, dx_3 = \delta_{\alpha\beta} \tag{2.35}$$

as in the case of the regular Sturm-Liouville problem. Although these functions belong to the Hilbert space under consideration $L_2(0, \infty; \mu)$, they do not build up a complete system within it, however. That is a typical situation for operators with mixed spectrum.

The improper eigenfunctions $U(\lambda, x_3)$ and the eigen packets or eigen differentials derived from them are related to the continuous part of the spectrum. These functions differ from each other in that the latter ones are elements of the Hilbert space, while the improper eigenfunctions do not belong to it. They fulfil only the weaker requirement for boundedness $\forall\, x_3, x_3 \in [0, \infty)$, and therefore the requirement 2) of (2.34) is violated [compare with SHEVCHENKO (1971a)]. Here we are dealing with two different possibilities of description for one and the same situation, in which we shall prefer the improper eigenfunctions. According to JÖRGENS and RELLICH (1976) it follows from the general theory of singular Sturm-Liouville operators that the improper eigenfunctions form an orthogonalized and normalized family of functions expressed by the orthonormality relation

$$\big(U(\lambda, x_3), U(\lambda', x_3)\big) = \int_0^\infty \mu U^*(\lambda, x_3)\, U(\lambda', x_3)\, dx_3 = \delta(\lambda - \lambda') \tag{2.36}$$

with the help of Dirac's δ-function. Further, the proper and improper eigenfunctions are orthogonal to each other, i.e.

$$\big(U(\lambda, x_3), {}^{\alpha}U\big) = 0. \tag{2.37}$$

From the spectral theory we obtain the important statement that the operator L is decomposable [see JÖRGENS and RELLICH (1976)], i.e. the eigenfunction system formed by ${}^{\alpha}U$ and $U(\lambda, x_3)$ is complete and any function $f(x_3)$, $f \in \vartheta(L)$ may be expanded in the manner

$$f(x_3) = \sum_{\alpha=1}^{M} f_\alpha {}^{\alpha}U(x_3) + \int_{-\infty}^{\Lambda} f_{(\lambda)} U(\lambda, x_3)\, d\lambda. \tag{2.38}$$

Here, f_α and $f_{(\lambda)}$ are the corresponding expansion coefficients, defined by

$$f_\alpha = \int_0^\infty \mu(\zeta)\, f(\zeta)\, {}^{\alpha}U(\zeta)\, d\zeta, \qquad f_{(\lambda)} = \int_0^\infty \mu(\zeta)\, f(\zeta)\, U(\lambda, \zeta)\, d\zeta. \tag{2.39}$$

The statements (2.35)—(2.39) result, as already mentioned, from the spectral theory of singular Sturm-Liouville operators. By using special assumptions they were also obtained by SHEVCHENKO (1979) for the infinite domain $x_3 \in (-\infty, \infty)$. The results given there hold their validity according to SHEVCHENKO (1978) also for the case $x_3 \in [0, \infty)$, if the boundary condition 3) from (2.34) is added. A detailed consideration of the Love wave operator was carried out by KAZI (1976) where the statements (2.35—2.39) are obtained by applying another method, described for example in FRIEDMAN's (1956) book. A Green's function is constructed and then integrated over a circle with an infinite radius in the complex plane. The integral is reduced to a sum of residues and a branch line integral. In this way a spectral expansion of the δ-function is obtained and the further required statements follow without difficulty.

Finally, let us refer to the fact that we have considered so far the eigenvalue problem (2.32) for fixed ω. However, sometimes it is useful [see LEVSHIN et al. (1986)] to consider also the case when k is fixed. Rearranging the eigenvalue equation (2.32) and using a modified operator $\tilde{L}$ and eigenvalue $\tilde{\lambda}$,

$$\tilde{L} = \frac{1}{\varrho}\left[-\frac{\mathrm{d}}{\mathrm{d}x_3}\left(\mu\frac{\mathrm{d}}{\mathrm{d}x_3}\right) + \mu k^2\right], \qquad \tilde{\lambda} = \omega^2, \tag{2.40}$$

we obtain in place of (2.32)

$$\tilde{L}U = \tilde{\lambda}U, \tag{2.41}$$

and the modified orthonormality relation is

$$\int_0^\infty \varrho\, {}^\alpha U^{*\beta} U \,\mathrm{d}x_3 = \delta_{\alpha\beta}. \tag{2.42}$$

That means, the weighting factor μ is replaced by the density ϱ. For our purposes, this representation is not convenient since for the discontinuity Snell's law requires fixed ω but not fixed k.

In the case of P-SV-motion (Rayleigh waves), a simple orthogonality relation holds for fixed k only:

$$\int_0^\infty \varrho\, {}^\alpha U_j^{*\beta} U_j \,\mathrm{d}x_3 = 0 \quad \text{for} \quad \alpha \neq \beta. \tag{2.43}$$

For fixed ω, the orthogonality relation is more complicated [cf. (3.27)]:

$$\int_0^\infty ({}^\alpha U_j{}^\beta S_{1j}^* - {}^\beta U_j^{*\alpha} S_{1j}) \,\mathrm{d}x_3 = 0 \quad \text{for} \quad \alpha \neq \beta. \tag{2.44}$$

In this case, the weighting factors enter into the x_3-dependences S_{1j} of the stress tensor components σ_{1j}.

3. The undisturbed wave-guide

3.1. The mode conception

Firstly, let us define the term wave-guide used here. A wave-guide is understood to be any structure capable of guiding energy parallel to its axis, at which the energy is restricted to the domain within, or in the vicinity of its surfaces. All wave-guides under consideration are covered by this definition. Although some fundamental ideas concerning the modes have already been developed in the introduction, we will give here a definition of the term mode. A wave-guide mode is a wave propagating along the wave-guide with well-defined phase and group velocity as well as cross-sectional intensity distribution (specifically the amplitude-depth-dependence) and polarization. The modes are the "characteristic waves" of the wave-guide structures, because their field vectors satisfy the homogeneous wave equation for all media of the wave-guide as well as the boundary and continuity conditions on the planes of discontinuity. Thus, the modes characterize the structure subject to its "resonances", and they do not depend on the sources of radiation exciting the wave-guide. Regardless of this conception, each mode propagating within a wave-guide can be interpreted as a superposition of plane waves making a well-defined angle with the wave-guide axis. This corresponds to the treatment of guided modes on the basis of ray optics.

3.2. The closed wave-guide

3.2.1. *General remarks*

Let us now consider an elastic, homogeneous wave-guide, whose Lamé parameter μ and density ϱ are constant over the whole cross-section (the other Lamé parameter is not relevant in this connection). The boundary conditions (2.23) are specified for two different wave-guide types as follows:

Wave-guide type I: $\nu_0 = \eta_0 = \mu$, $\nu_1 = \eta_1 = \varepsilon$, i.e.

$$\begin{aligned} \mu U'(0) + \varepsilon U(0) &= 0, \\ \mu U'(H) + \varepsilon U(H) &= 0, \end{aligned} \tag{3.1}$$

Wave-guide type II: $\nu_0 = \eta_0 = \mu$, $\nu_1 = \varepsilon$, $\eta_1 = 0$, i.e.

$$\begin{aligned} &\mu U'(0) + \varepsilon U(0) = 0, \\ &U'(H) = 0. \end{aligned} \tag{3.2}$$

The wave-guide type I was selected because its eigenvalues and eigenfunctions are very easily attainable. On the other hand, for the subsequent consideration of discontinuities, the wave-guide type II allows an analytical solution.

SHEVCHENKO (1971a) succeeded in obtaining some general statements about the structure of eigenvalues and eigenfunctions of an inhomogeneous (in x_3-direction), dielectric wave-guide without solving the corresponding eigenvalue problem directly. As a generalization of the case μ, ϱ = const these considerations may also be applied for the more complicated case $\mu = \mu(x_3)$, $\varrho = \varrho(x_3)$, because Shevchenko's eigenvalue equation changes into ours by substituting the quantities ε, k^2, and $\varkappa^2$ (using the nomenclature of Shevchenko) by $1/\mu$, $\dfrac{\varrho\omega^2}{\mu(1-\mu)}$, and $-\lambda$.

3.2.2. *The eigenspectrum*

3.2.2.1. Wave-guide type I

The differential equation (2.19) for the amplitude-depth-dependences of the separation statement

$$u(x_1, x_3, t) = U(x_3)\, e^{i(kx_1 - \omega t)} \tag{3.3}$$

may be written for the considered wave-guide by using the abbreviations

$$\gamma^2 = k_\beta^2 - k^2, \quad k_\beta^2 = \frac{\omega^2}{\beta^2} = \frac{\omega^2\varrho}{\mu} \tag{3.4}$$

as

$$U'' + \gamma^2 U = 0. \tag{3.5}$$

If the solution is sought as

$$U = C_1 e^{i\gamma x_3} + C_2 e^{-i\gamma x_3}, \tag{3.6}$$

the boundary conditions (3.1) require the vanishing of expression Δ_I,

$$\Delta_I = (e^{-i\gamma H} - e^{i\gamma H})\,(\mu^2\gamma^2 + \varepsilon^2). \tag{3.7}$$

By setting the first factor equal to zero the real γ-values

$$'\gamma = \pm\frac{\nu\pi}{H}, \quad \nu = 1, 2, \ldots \tag{3.8}$$

are obtained, by doing so with the second one, however we obtain the imaginary value

$$^0\gamma = \pm i \frac{\varepsilon}{\mu}. \tag{3.9}$$

By combining (2.21), (3.4), (3.8), and (3.9) the discrete, real eigenvalues λ_ν are obtained as

$$\lambda_0 = k_\beta^2 + \frac{\varepsilon^2}{\mu^2}, \quad \lambda_1 = k_\beta^2 - \frac{\pi^2}{H^2}, \quad \lambda_2 = k_\beta^2 - \frac{4\pi^2}{H^2}, \ldots \tag{3.10}$$

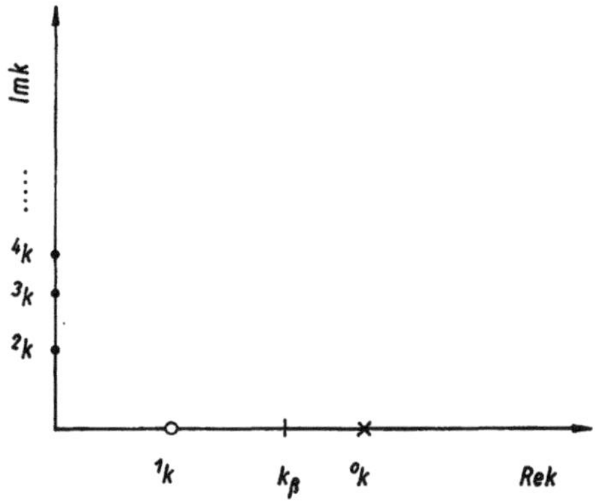

Fig. 7. The eigenvalues in the complex k-plane

In the following, let us consider the wave numbers $^\sigma k = \sqrt{\lambda_\sigma}$, $\sigma = 0, 1, 2, \ldots$, themselves as eigenvalues, i.e.,

$$^0k = \sqrt{k_\beta^2 + \frac{\varepsilon^2}{\mu^2}}, \quad ^\nu k = \sqrt{k_\beta^2 - \frac{\nu^2\pi^2}{H^2}}, \quad \nu = 1, 2, \ldots \tag{3.11}$$

Depending on whether the value of $\nu\pi/H$ is less or greater than k_β, is $^\nu k$ real or imaginary. In Fig. 7 the values $^\nu k$ are shown in the complex k-plane.

The modes of the wave-guide are related to these eigenvalues. To be more precise, there are a so-called "slow mode" with a real k, $k > k_\beta$ (corresponding to ^0k), a finite number M_s-depending on the period T — of "fast modes" with real eigenvalues in the domain $0 \leq k \leq k_\beta$ and an infinite number of modes with imaginary eigenvalues (wave numbers). The modes with real wave numbers are the actual "normal" modes of the wave-guide, which propagate with phase velocites $^\sigma c = \omega/^\sigma k$ ($\sigma = 0, 1, 2, \ldots, M_s$) into x_1-direction. For each fast mode a limit period T_ν,

$$T_\nu = \frac{2H}{\nu\beta}, \quad \nu = 1, 2, \ldots, M_s, \tag{3.12}$$

exists, above which the mode cannot propagate (cut-off). Only the slow mode has no upper limit. The relations (3.11) implicitely contain the dispersion relations, i.e., the dependence of the phase velocities on the period. They are obtained as (see Fig. 8)

$$^0c = \frac{1}{\sqrt{\frac{1}{\beta^2} + \left(\frac{\varepsilon T}{2\pi\mu}\right)^2}}, \quad ^\nu c = \frac{1}{\sqrt{\frac{1}{\beta^2} - \left(\frac{\nu T}{2H}\right)^2}}, \quad \nu = 1, \ldots, M_s. \tag{3.13}$$

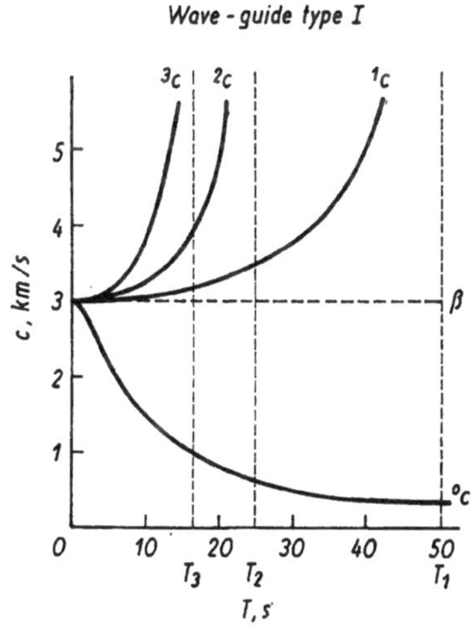

Fig. 8. Dispersion curves of the slow mode and the first three fast modes for the wave-guide type I

The changing character of the dispersion curve of the slow mode is obvious in Fig. 8: whereas the fast modes exhibit normal dispersion, we can see anomalous dispersion at the slow mode.

Because there is only a finite number of modes propagating into x_1-direction, they cannot form a complete eigenfunction system. This is connected with the fact that only a finite number of positive eigenvalues λ_ν exist for the corresponding eigenvalue problem according to section 2.2. In addition, there are an infinite number of modes with imaginary wave numbers which are attenuated in x_1-direction, which cannot propagate. They are needed for the subsequent treatment of discontinuities to describe the fine structure of the scattered field in the vicinity of discontinuities.

3.2.2.2. Wave-guide type II

We carry out here an analogous procedure as for type I. Instead of (3.7) now the expression Δ_{II}, defined by

$$\Delta_{\text{II}} = \tan \gamma H + \frac{\varepsilon}{\mu\gamma}, \tag{3.14}$$

has to vanish. This leads to the transcendental equation

$$\tan {}^{\nu}\gamma H + \frac{\varepsilon}{\mu\,{}^{\nu}\gamma} = 0 \tag{3.15}$$

with an infinite number of real roots ${}^{\nu}\gamma$, $\nu = 1, 2, \ldots$, corresponding to the multi-valuedness of the tangent function, and a purely imaginary root ${}^{0}\gamma = \mathrm{i}|{}^{0}\gamma|$, which appears as a solution of the modified equation (3.15)

$$\tanh |{}^{0}\gamma|\, H - \frac{\varepsilon}{\mu\,|{}^{0}\gamma|} = 0. \tag{3.16}$$

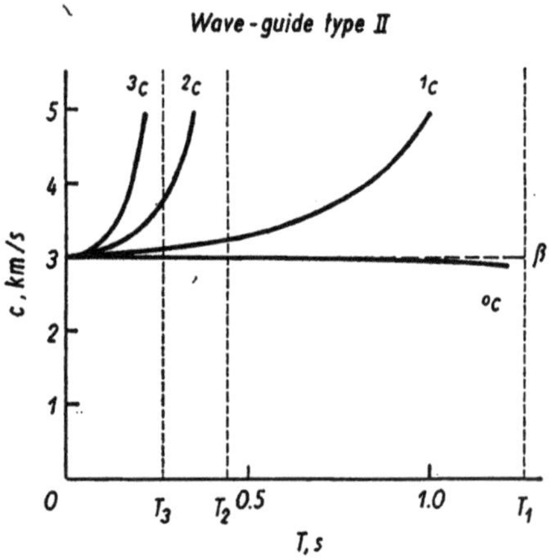

Fig. 9. Dispersion curves of the slow mode and the first three fast modes for the wave-guide type II

The eigenvalues follow analogously to (3.11) as

$$^{0}k = \sqrt{k_\beta^2 + {}^{0}\gamma^2}, \quad {}^{\nu}k = \sqrt{k_\beta^2 - {}^{\nu}\gamma^2}, \quad \nu = 1, 2, \ldots \tag{3.17}$$

That means we have here the same structure as for the wave-guide type I: slow mode — fast modes — attenuated modes, but the eigenvalues cannot be determined without solving a transcendental equation. In Fig. 9 the dispersion curves are plotted with the same parameters as for type I.

For this configuration the dispersion curves of the slow mode hardly differ from each other, but the upper limit periods of the fast modes are greatly shifted in the direction of lesser periods.

3.2.3. The eigenfunctions

3.2.3.1. Wave-guide type I

The integration constants in (3.6) are specified by the boundary conditions (3.1) and the eigenfunctions are obtained as

$$ {}^{\sigma}U(x_3) = {}^{\sigma}C\left(e^{i^{\sigma}\gamma x_3} - \frac{\varepsilon + i^{\sigma}\gamma\mu}{\varepsilon - i^{\sigma}\gamma\mu} e^{-i^{\sigma}\gamma x_3}\right), \quad \sigma = 0, 1, 2, \ldots, \tag{3.18} $$

where ${}^{\sigma}C$ is an amplitude factor. By inserting (3.9) into (3.18), at which the positive sign only leads to a physically significant solution, the amplitude-depth-dependence of the slow mode ${}^{0}U(x_3)$ follows as

$$ {}^{0}U(x_3) = {}^{0}C\, e^{-\frac{\varepsilon}{\mu} x_3}, \tag{3.19} $$

being exceptionally simple. Substituting (3.8) into (3.18), the analogous expression for the fast modes is obtained as

$$ {}^{\nu}U(x_3) = {}^{\nu}C\left(e^{i\frac{\nu\pi}{H} x_3} - \frac{\varepsilon + i\frac{\nu\pi}{H}\mu}{\varepsilon - i\frac{\nu\pi}{H}\mu} e^{-i\frac{\nu\pi}{H} x_3}\right), \quad \nu = 1, 2, \ldots \tag{3.20} $$

Here it is unimportant whether the positive or negative value of ${}^{\nu}\gamma$ is used. In Fig. 10 $\mathrm{Re}\,{}^{\sigma}U(x_3)$ $(\sigma = 0, 1, 2, 3)$ is plotted for a special model with the same parameters as in Fig. 8.

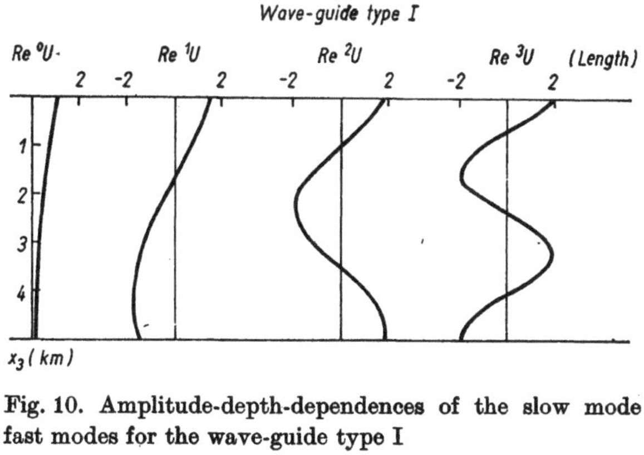

Fig. 10. Amplitude-depth-dependences of the slow mode and of three fast modes for the wave-guide type I

We can clearly recognize the "surface-wave-like" character of the slow mode.

For the sake of completeness the full expressions for the modes will be quoted here once more according to (3.3)

$$
\begin{aligned}
{}^0u(x_1, x_3, t) &= {}^0C\, \mathrm{e}^{-\frac{\varepsilon}{\mu}x_3}\, \mathrm{e}^{\pm \mathrm{i}\sqrt{k_\beta^2 + \frac{\varepsilon^2}{\mu^2}}\,x_1}\, \mathrm{e}^{-\mathrm{i}\omega t}, \\
{}_\nu u(x_1, x_3, t) &= {}^\nu C \left(\mathrm{e}^{\mathrm{i}\frac{\nu\pi}{H}x_3} - \frac{\varepsilon + \mathrm{i}\frac{\nu\pi}{H}\mu}{\varepsilon - \mathrm{i}\frac{\nu\pi}{H}\mu}\, \mathrm{e}^{-\mathrm{i}\frac{\nu\pi}{H}x_3} \right) \\
&\quad \times \mathrm{e}^{\pm \mathrm{i}\sqrt{k_\beta^2 - \frac{\nu^2\pi^2}{H^2}}\,x_1}\, \mathrm{e}^{-\mathrm{i}\omega t}, \quad \nu = 1, 2, \ldots
\end{aligned}
\tag{3.21}
$$

In the exponents, the two signs occur due to the ambiguity of the roots (3.11). The positive sign corresponds to modes propagating into x_1-direction if the wave numbers are real, and to modes attenuated in x_1-direction if they are imaginary. For the negative sign the same holds for the direction $-x_1$.

3.2.3.2. Wave-guide type II

The eigenfunctions again take the form (3.18), at which the corresponding γ-values are solutions of the equations (3.15) and (3.16), respectively. For the same parameters as in Fig. 10 Re ${}^\sigma U(x_3)$ ($\sigma = 0, 1, 2, 3$) is plotted in Fig. 11.

While the amplitude-depth-dependences of the slow modes hardly differ from each other there are remarkable differences for the fast ones.

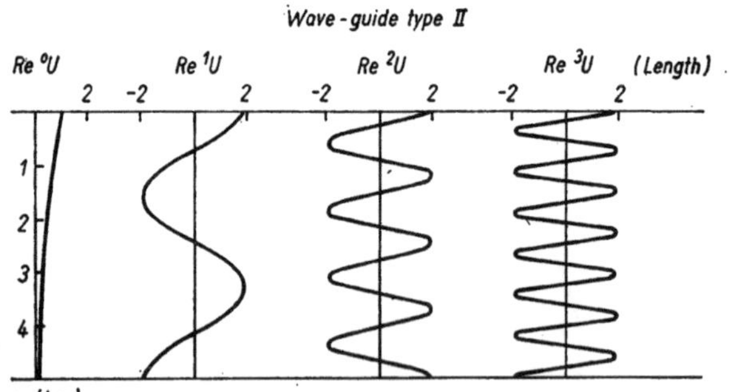

Fig. 11. Amplitude-depth-dependences of the slow mode and of three fast modes for the wave-guide type II

3.2.4. *Orthogonality and normalization*

The orthogonality both of the slow mode with the fast modes and of the fast modes one with another follows from (2.29),

$$\begin{aligned} &\int_0^H \mu^0 U^\nu U^* \,\mathrm{d}x_3 = 0, \qquad \nu = 1, 2, \ldots, \\ &\int_0^H \mu^\nu U^\mu U^* \,\mathrm{d}x_3 = 0, \qquad \nu, \mu = 1, 2, \ldots, \qquad \nu \neq \mu, \end{aligned} \tag{3.22}$$

and this can be confirmed immediately by evaluating the expressions (3.18), (3.19), and (3.20). The weight factor μ from the scalar product [see (2.25)] does not necessarily have to be taken into account, because it is a constant in our case. Nevertheless, it will be a contributing factor for the normalization. Further, it should be pointed out that analogously to the scalar product, the complex conjugate is used in formulating the orthogonality in (3.22). It refers to the so-called "energetic" orthogonality. Shevchenko (1971a) as well as Marcuse (1974) make a distinction between this and the "usual" orthogonality (without using the complex conjugate). The energetic orthogonality (or power orthogonality) has the physical consequence that the power carried by the wave-guide field is simply the sum of the powers carried by all the modes. This is not a trivial statement. For instance, in lossy dielectric wave-guides the power orthogonality does not hold any longer. However, the orthogonality in its usual sense applies. This mode orthogonality is more general than the power orthogonality. In all cases under consideration here, both kinds of orthogonality relations hold.

If in (3.22) both factors have the same mode index, this relation can be used for normalization, that is

$$\int_0^H \mu^0 U^0 U^* \,\mathrm{d}x_3 = \frac{{}^0C^2 \mu}{2\,|{}^0\gamma|} \left(1 - \mathrm{e}^{-2|{}^0\gamma| H}\right) \quad \text{and} \tag{3.23}$$

$$\int_0^H \mu^\nu U^\nu U^* \,\mathrm{d}x_3 = \mu^\nu C^2 \left[2H + \frac{2\mu\varepsilon(\cos 2^\nu\gamma H - 1) + \left({}^\nu\gamma\mu^2 - \frac{\varepsilon^2}{{}^\nu\gamma}\right) \sin 2^\nu\gamma H}{\varepsilon^2 + {}^\nu\gamma^2\mu^2} \right], \quad \nu = 1, 2, \ldots \tag{3.24}$$

In this way, with the corresponding ${}^\nu\gamma$-values ($\nu = 1, 2, \ldots$) for wave-guides of type I or II the orthonormality relations of the fast modes are obtained as

$$\int_0^H \mu^\nu U^\mu U^* \, dx_3 = \mu^\nu C^2 \delta_{\mu\nu} \left[2H + \frac{2\mu\varepsilon(\cos 2^\nu\gamma H - 1) + \left({}^\nu\gamma\mu^2 - \frac{\varepsilon^2}{{}^\nu\gamma}\right) \sin 2^\nu\gamma H}{\varepsilon^2 + {}^\nu\gamma^2\mu^2} \right],$$

$$\nu, \mu = 1, 2, \ldots \tag{3.25}$$

Particularly noteworthy is the fact that for wave-guide type I, equation (3.25) has the very simple form

$$\int_0^H \mu^\nu U^\mu U^* \, dx_3 = 2\mu H^\nu C^2 \, \delta_{\mu\nu}, \qquad \mu, \nu = 1, 2, \ldots \tag{3.26}$$

Without loss of generality we have assumed, that the mode amplitudes ${}^\sigma C$ ($\sigma = 0, 1, 2, \ldots$) are real. For normalization, it is suitable to use the formula for the energy flux ${}^\sigma S$ of mode σ into x_1-direction striking an infinitely deep strip of unit width in the x_2,x_3-plane per unit time [see for instance formula (41) in MALISCHEWSKY (1973)]:

$${}^\sigma S = \frac{i\omega}{4} \int_0^H ({}^\sigma U_j {}^\sigma S^*_{1j} - {}^\sigma U^{*\sigma}_j S_{1j}) \, dx_3. \tag{3.27}$$

This formula is considerably simplified for shear wave propagation:

$${}^\sigma S = \frac{i\omega}{4} \int_0^H ({}^\sigma U {}^\sigma S^*_{12} - {}^\sigma U^{*\sigma} S_{12}) \, dx_3 = \frac{\omega}{4} ({}^\sigma k + {}^\sigma k^*) \int_0^H \mu^\sigma U^\sigma U^* \, dx_3. \tag{3.28}$$

As usual the quantities ${}^\sigma S_{ij} = {}^\sigma S_{ij}(x_3)$ are the x_3-dependences of the stress tensor components for mode σ. As expected we recognize from (3.28), that ${}^\sigma S$ vanishes for modes with imaginary ${}^\sigma k$ (i.e. for modes being attenuated in x_1-direction). By normalizing the slow mode in such a manner that it is transporting unit energy, and by using (3.9), (3.23), and (3.28), the amplitude factor ${}^0C$ is obtained for wave-guide type I as

$${}^0C = \frac{2}{\mu} \sqrt{\frac{\varepsilon}{{}^0k\omega(1 - e^{-2\varepsilon H/\mu})}} \tag{3.29}$$

and for wave-guide type II as

$${}^0C = 2 \sqrt{\frac{{}^0\gamma}{{}^0k\omega\mu(1 - e^{-2{}^0\gamma H})}}. \tag{3.30}$$

It would be possible to normalize the unattenuated fast modes in the same way (unit energy transport). But to achieve better correspondence with the open wave-guide we will not do so. On the contrary, let us normalize the fast modes altogether, so that the orthonormality relation is

$$\int_0^H \mu^\nu U^\mu U^* \, \mathrm{d}x_3 = \mu\delta_{\mu\nu}, \qquad \nu, \mu = 1, 2, \ldots, \tag{3.31}$$

and therefore, by using (3.26) ${}^\nu C$ is given by

$${}^\nu C = \frac{1}{\sqrt{2H}}, \quad \nu = 1, 2, \ldots, \tag{3.32}$$

for wave-guide type I and according to (3.25) a more complicated relation is obtained for type II.

3.3. The open wave-guide

3.3.1. General remarks

In what follows, the limiting process from the closed to the open wave-guide is performed by increasing the thickness H to infinity. As in section 3.2. we again have in view an elastic, homogeneous wave-guide. In many cases and in some respects the closed wave guide allows a much simpler treatment than the open one. In certain situations, however, just the opposite is valid, as will be seen later. The limiting process mentioned above was performed by a number of authors in the one as well as in the other direction with a view to the numerical treatment of wave-guide problems [see for example Shevchenko (1971a), Brooke and Kharadly (1976), Pitts et al. (1976), and Yoneyama and Nishida (1981)].

At the surface of the open wave-guide let the impedance-like boundary condition

$$\mu U'(0) + \varepsilon U(0) = 0 \tag{3.33}$$

hold, analogously to (3.1). For the open wave-guide, the second boundary condition for $x_3 = H$ is replaced by a condition which controls the behaviour of the solutions in infinity. For the present, we shall merely set up the physically reasonable requirement that $U(x_3)$ is a member of the space $\mathrm{C}_2(0, \infty)$ of twice continuously differentiable functions in the interval $[0, \infty)$, which are consequently bounded:

$$U(x_3) \in \mathrm{C}_2(0, \infty). \tag{3.34}$$

Taking this assumption as a basis the eigenspectrum and the eigenfunctions of the open wave-guide are investigated in the following.

3.3.2. *The eigenspectrum*

The formulas (3.3)—(3.6) hold in the same way for the open wave-guide, too. The boundary condition (3.33) leads to the equation

$$C_1(\varepsilon + i\gamma\mu) + C_2(\varepsilon - i\gamma\mu) = 0, \tag{3.35}$$

where γ is arbitrary for the present, while nevertheless being restricted by the additional requirement (3.34). Obviously all real γ and the imaginary value

$$^0\gamma = \pm i \frac{\varepsilon}{\mu} \tag{3.36}$$

are allowed. The real γ form the continuous spectrum which covers the interval $[0, k_\beta]$ of the real axis and the positive imaginary semi-axis in the complex k-plane. Owing to the ambiguity of the square root in (3.4), the same is true analogously for the interval $[0, -k_\beta]$ and the negative imaginary semi-axis, respectively. In accordance with (3.11) the (discrete) imaginary γ-value leads to

$$^0k = \sqrt{k_\beta^2 + \frac{\varepsilon^2}{\mu^2}}, \tag{3.37}$$

so that the distribution of eigenvalues shown in Fig. 12 is obtained.

In comparison with Fig. 7 it can be recognized that the discrete values $^{\nu}k$

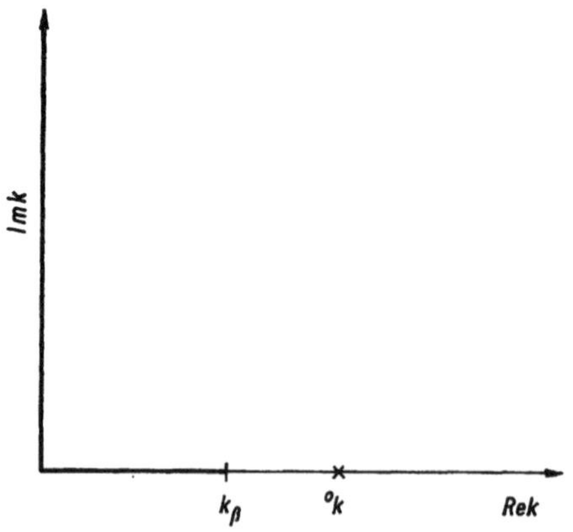

Fig. 12. The eigenspectrum of the open wave-guide

($\nu = 1, 2, \ldots$) have moved together to form a continuum, while the slow mode has kept its place and is from now on forming the surface wave in the open wave-guide. The dispersion relation is the same as for the closed wave-guide. The continuum of eigenvalues corresponds to the so-called radiation modes or body waves, which can propagate into x_1-direction for real k and which are attenuated for imaginary k. The latter cause a certain energy storage in the vicinity of discontinuities, disturbing bodies, and so on.

If the wave-guide is inhomogeneous in x_3-direction, a peculiarity of the spectrum may occur, to which the reader's attention is drawn in the following separate section.

3.3.3. *Leaking modes*

Obviously, there are no complex eigenvalues for an homogeneous wave-guide. In principle, the same statement holds for the inhomogeneous wave-guide too, because it is certainly a singular but a self-adjoint boundary value problem. Nevertheless, an essential peculiarity may occur. Some fundamental details of this are discussed here.

The self-adjointness of the boundary value problem does not depend on the inhomogeneity of the open wave-guide, e.g. it is independent of the number of layers. But quoting MOLOTKOV (1974), the boundedness of solutions in the whole domain is a necessary requirement for the self-adjointness of the problem. By dropping the "boundary condition" (3.34), i.e. by searching for so-called non-physical solutions, we are led automatically to a non-self-adjoint operator with complex eigenvalues. The corresponding solutions are denoted by leaking modes, because energy is leaked into the half-space during the propagation into direction x_1. Incidently, there is also another approach to leaking modes carried out by proceeding from an inhomogeneous wave-guide being homogeneous from the depth $x_3 = \bar{H}$. The dispersion relation can then be interpreted as a boundary condition for $x_3 = \bar{H}$ and a regular but non-self-adjoint boundary value problem is obtained with partially complex eigenvalues [see for instance SHEVCHENKO (1971 b)]. It is true that by this means the eigenvalues are discretized, but the completeness is lost.

By using the simplest elastic model in which leaking modes are present, let us demonstrate the occurrence of complex eigenvalues. Our model consists of an elastic layer with thickness H, shear wave velocity β_1, and mass density ϱ_1, in welded contact with an elastic half-space with shear wave velocity β_2, where $\beta_2 > \beta_1$, and density ϱ_2. The solution is sought in the layer and the half-space, respectively, as

$$U_1 = C_1 \, e^{i\gamma_1 x_3} + C_2 \, e^{-i\gamma_1 x_3} \tag{3.38}$$

and

$$U_2 = C_3 \, e^{i\gamma_2 x_3} + C_4 \, e^{-i\gamma_2 x_3}, \tag{3.39}$$

where analogous to (3.4) the abbreviations

$$\gamma_1 = \sqrt{k_{\beta_1}^2 - k^2}, \qquad \gamma_2 = \sqrt{k_{\beta_2}^2 - k^2};$$
$$k_{\beta_1}^2 = \frac{\omega^2}{\beta_1^2} = \frac{\omega^2 \varrho_1}{\mu_1}, \qquad k_{\beta_2}^2 = \frac{\omega^2}{\beta_2^2} = \frac{\omega^2 \varrho_2}{\mu_2} \tag{3.40}$$

are used. To exclude the radiation modes for a short time, the condition (3.34) has to be replaced by the more rigorous one

$$U(x_3) \in \mathbf{L}_2, \tag{3.41}$$

being equivalent with $U(\infty) = 0$. For this reason $\operatorname{Re} \gamma_2 = 0$ must hold, and by assuming $\operatorname{Im} \gamma_2 > 0$, $C_4 = 0$. Hence, the boundary condition (3.33) and the continuity conditions according to (2.24) yield the system of equations

$$(\mathrm{i}\mu_1\gamma_1 + \varepsilon)\, C_1 - (\mathrm{i}\mu_1\gamma_1 - \varepsilon)\, C_2 = 0,$$
$$\mathrm{e}^{\mathrm{i}\gamma_1 H} C_1 + \mathrm{e}^{-\mathrm{i}\gamma_1 H} C_2 - \mathrm{e}^{\mathrm{i}\gamma_2 H} C_3 = 0, \tag{3.42}$$
$$\mathrm{e}^{\mathrm{i}\gamma_1 H} C_1 - \mathrm{e}^{-\mathrm{i}\gamma_1 H} C_2 - \frac{\mu_2\gamma_2}{\mu_1\gamma_1}\, \mathrm{e}^{\mathrm{i}\gamma_2 H} C_3 = 0,$$

whose consistency condition yields the dispersion relation in the form

$$\frac{\mu_1\gamma_1 + \mu_2\gamma_2}{\mu_1\gamma_1 - \mu_2\gamma_2} = \frac{\mathrm{i}\mu_1\gamma_1 - \varepsilon}{\mathrm{i}\mu_1\gamma_1 + \varepsilon}\, \mathrm{e}^{2\mathrm{i}\gamma_1 H}. \tag{3.43}$$

By setting $\varepsilon = 0$, the well-known dispersion relation of Love waves for a one-layer model follows:

$$\frac{\mu_1\gamma_1 + \mu_2\gamma_2}{\mu_1\gamma_1 - \mu_2\gamma_2} = \mathrm{e}^{2\mathrm{i}\gamma_1 H}. \tag{3.44}$$

In the following let us consider precisely this case for the sake of simplicity. The two Riemannian sheets, which are necessary for the definition of a complex $k(\omega)$, are denoted by (+) and (−) in accordance with the sign of $\operatorname{Im} \gamma_2$. The only branch point is k_{β_2}. It is well-known that a finite number of real roots ${}^{\nu}k$ exist on the (+)-sheet for a given ω, which are located in the interval

$$k_{\beta_2} < {}^{\nu}k < k_{\beta_1}. \tag{3.45}$$

There are no complex roots on the (+)-sheet. These are located on the (−)-sheet and correspond to the leaking modes. To investigate them more thoroughly it is advisable to start from small ω according to WATSON (1970). For the limit $\omega \to 0$, $\gamma_1 \to \mathrm{i}k$ and $\gamma_2 \to -\mathrm{i}k$ is obtained due to (3.40). Substituting these limits into (3.44) the low-frequency approximation

$$\frac{\mu_1 - \mu_2}{\mu_1 + \mu_2} = \mathrm{e}^{-2kH} = \mathrm{e}^{-2(\xi + \mathrm{i}\zeta)} = \mathrm{e}^{-2\xi}(\cos 2\zeta - \mathrm{i} \sin 2\zeta), \quad kH \overset{!}{=} \xi + \mathrm{i}\zeta \tag{3.46}$$

is obtained. By separating into real and imaginary parts it follows that

$$\begin{aligned} \zeta &= \pm \frac{(2j-1)}{2}\pi \quad \text{for} \quad \mu_1 < \mu_2 \quad \text{and} \\ \zeta &= \pm j\pi \quad \text{for} \quad \mu_1 > \mu_2, \quad j = 1, 2, \ldots, \end{aligned} \tag{3.47}$$

and

$$\begin{aligned} \xi &= \operatorname{artanh} \frac{\mu_1}{\mu_2} \quad \text{for} \quad \mu_1 < \mu_2 \quad \text{and} \\ \xi &= \operatorname{artanh} \frac{\mu_2}{\mu_1} \quad \text{for} \quad \mu_1 > \mu_2. \end{aligned} \tag{3.48}$$

Consequently, the low-frequency eigenvalues on the (—)-sheet appear as complex conjugate pairs at the points

$$\begin{aligned} {}^{\nu}k &= \frac{1}{H}\left[\operatorname{artanh} \frac{\mu_1}{\mu_2} \pm \mathrm{i}\, \frac{(2\nu-1)}{2}\pi\right] \quad \text{for} \quad \mu_1 < \mu_2 \quad \text{and} \\ {}^{\nu}k &= \frac{1}{H}\left[\operatorname{artanh} \frac{\mu_2}{\mu_1} \pm \mathrm{i}\nu\pi\right] \quad \text{for} \quad \mu_1 > \mu_2, \quad \nu = 1, 2, \ldots \end{aligned} \tag{3.49}$$

For an impedance surface [see (3.43)], similar results cannot be obtained so rapidly, because a coupled non-linear system of equations for ξ and ζ has to be solved. According to WATSON (1970) the dispersion curves of Love waves may be computed by following roots in the k-plane, using (3.47) as the starting point for each mode.

However, it is also possible to obtain starting values with $\omega \neq 0$, necessary for the application of numerical methods, by using a semi-analytical method which was employed by MARCUSE (1974) for the investigation of leaking modes in optical wave-guides. In this way, a deeper insight into the behaviour of complex roots is obtained, and that is the reason why we want to demonstrate this method for leaking modes of Love waves in some detail [see MALISCHEWSKY (1985)]. By using the abbreviations

$$x = \gamma_1 H, \quad m = \frac{\varrho_2\beta_2^2}{\varrho_1\beta_1^2}, \quad \tau = \frac{\omega}{\omega_c}, \quad \omega_c = \frac{\beta_1\pi}{(1-\beta_1^2/\beta_2^2)^{1/2}\cdot H} \tag{3.50}$$

the dispersion relation of Love waves in the one-layer case (3.44) can also be written as

$$m \operatorname{sgn}(\operatorname{Im} \gamma_2)\, \frac{(\tau^2\pi^2 - x^2)^{1/2}}{x} = \tan x, \tag{3.51}$$

where ω_c is the cut-off frequency of the first higher mode, and τ characterizes the deviation of the frequency from ω_c. By assuming $\operatorname{Im} \gamma_2 > 0$ the roots x_ν,

$\nu = 0, 1, 2, \ldots$, of the transcendental equation (3.51) lead to the real eigenvalues of the normal modes according to

$$^{\nu}k = (k_{\beta_1}^2 - x_\nu^2/H^2)^{1/2}, \qquad \nu = 0, 1, 2, \ldots \tag{3.52}$$

The fundamental mode is equivalent to $\nu = 0$. These real roots are the only ones located on the sheet $\operatorname{Im} \gamma_2 > 0$ of the Riemannian surface. As mentioned above, complex roots have to be searched for on the sheet $\operatorname{Im} \gamma_2 < 0$. The corresponding eigenfunctions grow exponentially as $x_3 \to \infty$ [see (3.39)]. In trying to find solutions of the equation

$$\tan x + m \frac{(\tau^2\pi^2 - x^2)^{1/2}}{x} = 0 \tag{3.53}$$

it is noted that there are two real roots in a certain range of τ, $\bar{\tau}_\nu < \tau \leqq \nu$, which represent the crossings of the curve $-(m/x)\,(\tau^2\pi^2 - x^2)^{1/2}$ with the particular branch of $\tan x$. These solutions are also unphysical, because the eigenfunctions are unbounded as $x_3 \to \infty$. One of them changes into a normal mode for $\tau > \nu$, and into a leaking mode for $\tau < \bar{\tau}_\nu$. For $\tau = \bar{\tau}_\nu$ the two real solutions mentioned above degenerate into one root only. This behaviour is shown in Fig. 13 for the first and second higher modes.

The parameters for the model under consideration are $\beta_1 = 3$ km/s, $\beta_2 = 4$ km/s, $\varrho_1 = 2$ g/cm^3, $\varrho_2 = 3$ g/cm^3, $H = 10$ km. For values of τ smaller than $\bar{\tau}_\nu$, there are no more real solutions. The value $\bar{\tau}_\nu$ at which the real unphysical solution of equation (3.53) just ceases to exist is of interest. We obtain the solution at this point by requiring that the tangents of the functions $\tan x$

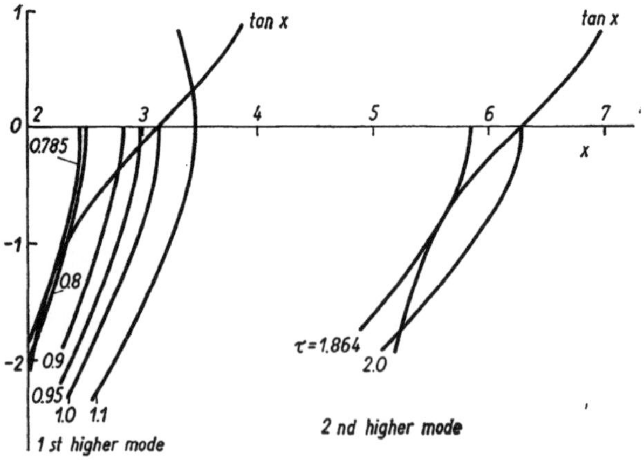

Fig. 13. Solutions of equation (3.51) for various values of τ

and $-(m/x)\,(\tau^2\pi^2 - x^2)^{1/2}$ are parallel and that the appropriate $\bar{x}_\nu$ still satisfies equation (3.53). After some transformations we find

$$\bar{\tau}_\nu = \frac{1}{\pi}\left[\frac{y_\nu^3(m^2-1)}{m-y_\nu}\right]^{1/2}, \quad \bar{x}_\nu = \left[\frac{m y_\nu^2(m y_\nu - 1)}{m - y_\nu}\right]^{1/2}, \quad \nu = 1, 2, \ldots, \tag{3.54}$$

where y_ν are the solutions of the transcendential equation

$$\left[\frac{m(m-y)}{my-1}\right]^{1/2} + \tan\left[\frac{my^2(my-1)}{m-y}\right]^{1/2} = 0. \tag{3.55}$$

Below the limit $\bar{\tau}_\nu$, the solutions of (3.53), if they exist at all, must be complex. The crucial point is the assumption that the solutions move continuously from real to complex values, so that their imaginary parts begin with small values. So it appears to be promising to apply a perturbation method, the basis of which is given in Marcuse (1974) (see Appendix 7.2.). Let

$$\tau^2 = \bar{\tau}^2 - \eta^2 \tag{3.56}$$

and

$$f_1(x) = \tan x, \quad f_2(x, \eta) = -m\,\frac{[(\bar{\tau}^2 - \eta^2)\,\pi^2 - x^2]^{1/2}}{x}, \tag{3.57}$$

so that $F(x, \eta) = f_1(x) - f_2(x, \eta) = 0$ is identical with equation (3.53) and

$$F(\bar{x}_\nu, 0) \equiv 0 \tag{3.58}$$

is valid. The function $f_2(x, \eta)$ was chosen in this manner to fulfil the necessary condition $\partial f_2(\bar{x}_\nu, 0)/\partial\eta = 0$ of the perturbation method. We obtain the complex approximate solutions in the range $\tau < \bar{\tau}_\nu$ as

$$x_\nu \approx \bar{x}_\nu \pm \mathrm{i}B_1^{(\nu)}\left(\bar{\tau}_\nu^2 - \frac{\omega^2}{\omega_c^2}\right)^{1/2} \tag{3.59}$$

with

$$\frac{1}{B_1^{(\nu)2}} = \frac{\bar{\tau}_\nu^2}{\bar{x}_\nu^2}\cdot\frac{3\bar{x}_\nu^2 - 2\bar{\tau}_\nu^2\pi^2}{\bar{\tau}_\nu^2\pi^2 - \bar{x}_\nu^2} - \frac{2\bar{x}_\nu}{m\pi^2}\,\frac{\sin\bar{x}_\nu}{\cos^3\bar{x}_\nu}\,(\bar{\tau}_\nu^2\pi^2 - \bar{x}_\nu^2)^{1/2}. \tag{3.60}$$

For the present model, the characteristic values concerning the first and second higher modes are given in Table 1.

Table 1. Characteristic values for the first and second higher modes

ν	y_ν	$\bar{x}_\nu$	$\bar{\tau}_\nu$	$B_1^{(\nu)}$
1	1.1498	2.1912	0.7877	0.9981
2	1.7352	5.5919	1.8637	0.7144

The complex eigenvalues (wave numbers) of the leaking modes are obtained by putting (3.59) into (3.52) as

$$^{\nu}k = (k_{\beta_1}^2 - \bar{x}_\nu^2/H^2)^{1/2} \pm \mathrm{i}\,\frac{\bar{x}_\nu B_1^{(\nu)}}{H^2(k_{\beta_1}^2 - \bar{x}_\nu^2/H^2)^{1/2}} + O(\eta^2), \qquad (3.61)$$

and the phase velocity is defined by

$$c_\nu = \omega/\mathrm{Re}\,{}^{\nu}k. \qquad (3.62)$$

Having clarified the fine structure in the vicinity of the limiting period of normal modes, we are now able to use the approximate leaking mode solutions, especially the critical point $\bar{x}_\nu$, as favourable starting values for a computer program. By applying the program CREG, originally used by Clayton

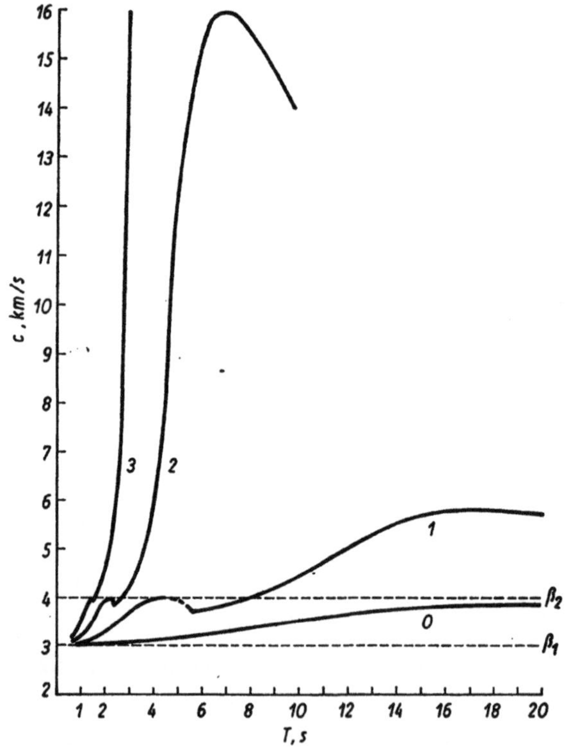

Fig. 14. Dispersion curves of the first three leaking modes, together with the normal mode

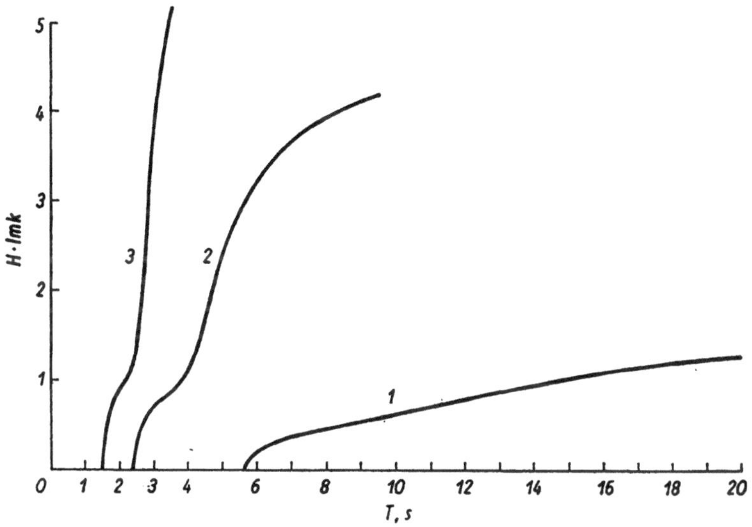

Fig. 15. The attenuation factor of the first three leaking modes versus period

and DERRICK (1977) for finding complex eigenvalues of wave equations, we succeed in determining zeros of the complex function (3.53). The basis of CREG is the well-known Regula Falsi technique, which can be modified for calculations in the complex plane. In Fig. 14 the phase velocity versus period is plotted for the first three leaking modes belonging to the corresponding normal modes together with the fundamental mode.

Fig. 15 shows the logarithm of the attenuation factor for each mode, i.e., $H \cdot \mathrm{Im}({}^{\nu}k)$ $(\nu = 1, 2, 3)$ versus period.

In Fig. 14, we note that there is no continuous transition from normal to leaking modes. So-called "open modes" [see BUDDEN (1961)] exist in this range (dashed lines), whose physical evidence cannot be discussed here. They correspond to one of the two real solutions existing on the (—)-sheet, which have already been mentioned. Further, the leaking modes start with a phase velocity value below the maximum shear wave velocity of the model (in our case β_2). This behaviour is contradictory to the interpretation of leaking modes as corresponding to a zig-zag path of rays in a wave-guide with under-critical reflections [see e.g. ALSOP et al. (1974)]. A more thorough consideration shows [see MALISCHEWSKY (1985)] that leaking modes are intrinsically connected with complex angles of incidence. Another point worth mentioning is the fact that, in conformity with RADOVICH and DE BREMAECKER (1974), all leaking Love-wave modes are highly damped with the exception of periods close to $\overline{T}_\nu = 2\pi/(\overline{\tau}_\nu \cdot \omega_c)$ (see Fig. 15). Finally let us state once more that a leaking mode is the analytic continuation of a normal mode below its cut-off frequency. It is an improper spectral mode solution of the corresponding equations of motion. These solutions, which are actually unphysical, may be regarded as accounting more efficiently for the normal mode continuous spectrum contribution in a bounded part of the wave-guide [see e.g. KAMEL and FELSEN (1981)].

Let us finish this section with the remark, that an homogeneous half-space with an impedance-like boundary condition on its surface is capable of guiding leaking modes with Rayleigh-kind motion [see VIKTOROV (1981)].

3.3.4. *The eigenfunctions*

In conformity with former considerations, the family of eigenfunctions of the open, homogeneous wave-guide under consideration consists of the discrete surface wave mode

$$ {}^0U(x_3) = {}^0C\,\mathrm{e}^{-\frac{\varepsilon}{\mu}x_3}, \tag{3.63} $$

having the same appearence as the corresponding slow mode of the closed wave-guide [compare with (3.19)], and of the continuum of radiation modes

$$ U(x_3, \gamma) = C\left(\mathrm{e}^{\mathrm{i}\gamma x_3} - \frac{\varepsilon + \mathrm{i}\gamma\mu}{\varepsilon - \mathrm{i}\gamma\mu}\,\mathrm{e}^{-\mathrm{i}\gamma x_3}\right), \quad |\gamma| \in [0, \infty). \tag{3.64} $$

In the following, let the parameter γ vary in the interval $0 \leqq \gamma < \infty$ only, because the variation in the interval $-\infty < \gamma \leqq 0$ does not yield any new function. Analogous to (3.21) the complete expressions for the modes are obtained as

$$ {}^0u(x_1, x_3, t) = {}^0C\,\mathrm{e}^{-\frac{\varepsilon}{\mu}x_3}\,\mathrm{e}^{\pm\mathrm{i}\sqrt{k_\beta^2 + \frac{\varepsilon^2}{\mu^2}}\,x_1}\,\mathrm{e}^{-\mathrm{i}\omega t}, \tag{3.65} $$

$$ \begin{aligned} u(x_1, x_3, \gamma, t) = C&\left(\mathrm{e}^{\mathrm{i}\gamma x_3} - \frac{\varepsilon + \mathrm{i}\gamma\mu}{\varepsilon - \mathrm{i}\gamma\mu}\,\mathrm{e}^{-\mathrm{i}\gamma x_3}\right)\\ &\times \mathrm{e}^{\pm\mathrm{i}\sqrt{k_\beta^2 + \gamma^2}\,x_1}\,\mathrm{e}^{-\mathrm{i}\omega t}, \qquad \gamma \in [0, \infty). \end{aligned} \tag{3.66} $$

One part of the spectrum of radiation modes can be interpreted in a physically intuitive manner. Let γ vary in the interval $[0, k_\beta]$ and let

$$ \gamma = k_\beta\,|\sin\bar{\varphi}| \quad \text{with} \quad 0^\circ \leqq \bar{\varphi} \leqq 180^\circ \tag{3.67} $$

then the second term in the bracket of (3.21) is an incident plane wave (body wave), and the first one is also a plane wave, reflected from the surface. The quantity $\bar{\varphi}$ is the angle between the wave number vector of the incident wave and the positive x_1-axis. It varies from 0° to 90° for radiation modes propagating into direction x_1, and from 180° to 90° for the opposite direction of propagation (see Fig. 16).

Obviously, within this domain of variability each angle of incidence is allowed and therefore the radiation modes form a continuum. From the mathematical point of view the reason for this fact is that the radiation modes consist of both incident and reflected waves. Thus, an inhomogeneous equation follows and there is no necessity for a determinant to vanish as in (3.7). The

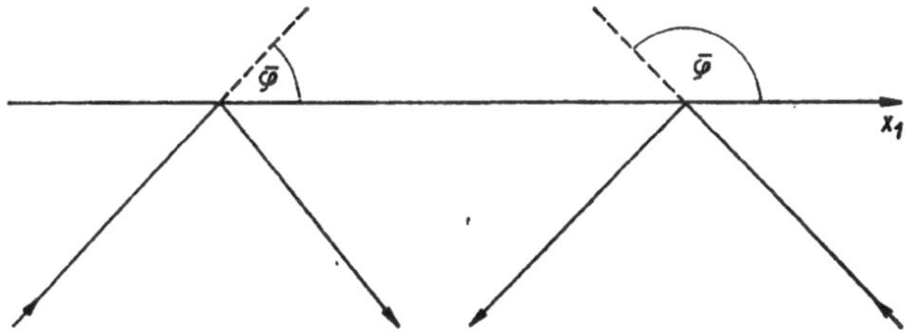

Fig. 16. Illustration of radiation modes

radiation modes represent standing waves concerning the x_3-direction [see (3.64)]. However, in the domain $k_\beta < \gamma < \infty$, the angle of incidence $\bar{\varphi}$ becomes imaginary according to (3.67) and, consequently, the physically intuitive interpretation of radiation modes fails. This part of the continuous spectrum is sometimes called reactive, as opposed to the radiative part [see Rozzi (1978)]. As already mentioned the reactive part is attenuated in the direction of propagation and it causes an energy storage in the vicinity of discontinuities. It is to be emphasized here that the mechanism of attenuation is based on the nature of these eigenfunctions and not, as in the case of leaking modes, on radiation into the half-space. Finally, in Table 2 a survey is given about the eigenfunctions as well as the spectrum of the open and the closed wave-guide.

Table 2. Correspondence between eigenfunctions and spectrum of the open and the closed wave-guide

	Open		Closed	
Wave number k	Mode	Spectrum	Mode	Spectrum
$k > k_\beta$	Surface wave	Discrete	Slow	Discrete
$0 \leq k \leq k_\beta$	Radiation	Continuous	Fast	Discrete
$i0 < k < i\infty$	Attenuated radiation	Continuous	Evanescent	Discrete
Complex	Leaking	Discrete	—	—

3.3.5. *Orthogonality and normalization*

Due to (2.36) and (2.37), orthogonality relations hold for the surface wave with the radiation modes (i.e. body waves) as well as for radiation modes one with another,

$$\int_0^\infty \mu^0 U(x_3)\, U^*(x_3, \gamma)\, dx_3 = 0,$$

$$\int_0^\infty \mu U^*(x_3, \gamma)\, U(x_3, \gamma')\, dx_3 = 0, \tag{3.68}$$

$$\gamma \neq \gamma' \;\&\; \gamma, \gamma' \in [0, \infty),$$

and this can also be confirmed directly by using the expressions (3.63) and (3.64). But the calculation concerning the second relation of (3.68) is not trivial and therefore we want to represent it in some detail in the course of normalization. For surface waves, the normalization is carried out by using (3.63) on the basis of relation [in analogy to (3.23)]

$$\int_0^\infty \mu^0 U^0 U^*\, dx_3 = \frac{{}^0C^2 \mu^2}{2\varepsilon}. \tag{3.69}$$

According to (2.36) the δ-function appears in the course of normalization of radiation modes and body waves, respectively, and that is why the integrals have to be calculated with great care. By using (3.64) we obtain

$$\int_0^\infty \mu U^*(x_3, \gamma)\, U(x_3, \gamma')\, dx_3$$

$$= C^*C\mu \left\{ \int_0^\infty e^{-i(\gamma-\gamma')x_3}\, dx_3 - \frac{\varepsilon + i\gamma'\mu}{\varepsilon - i\gamma'\mu} \int_0^\infty e^{-i(\gamma+\gamma')x_3}\, dx_3 \right. \tag{3.70}$$

$$\left. - \frac{\varepsilon - i\gamma\mu}{\varepsilon + i\gamma\mu} \int_0^\infty e^{i(\gamma+\gamma')x_3}\, dx_3 + \frac{\varepsilon - i\gamma\mu}{\varepsilon + i\gamma\mu} \cdot \frac{\varepsilon + i\gamma'\mu}{\varepsilon - i\gamma'\mu} \int_0^\infty e^{i(\gamma-\gamma')x_3}\, dx_3 \right\}.$$

It is well-known, that

$$\frac{1}{2\pi} \int_{-\infty}^\infty e^{i(\varrho-\sigma)x_3}\, dx_3 = \delta(\varrho - \sigma) \tag{3.71}$$

is valid, but for the interval [0, ∞) another relation

$$\int_0^\infty e^{i(\varrho-\sigma)x_3}\,dx_3 = \pi\delta(\varrho-\sigma) + iP\frac{1}{\varrho-\sigma} \tag{3.72}$$

holds, where P means the principal value which is only defined under the integral sign as

$$\int_{-a}^{b} P\frac{1}{x}\,f(x)\,dx = \lim_{\tau\to 0}\left[\int_{-a}^{-\tau}\frac{f(x)}{x}\,dx + \int_{+\tau}^{b}\frac{f(x)}{x}\,dx\right]. \tag{3.73}$$

Using (3.72) it follows from (3.70) that

$$\begin{aligned}\int_0^\infty &\mu U^*(x_3,\gamma)\,U(x_3,\gamma')\,dx_3\\ &= C^*C\mu\left\{\pi\delta(\gamma'-\gamma) + iP\frac{1}{\gamma'-\gamma}\right.\\ &\quad - \frac{\varepsilon+i\gamma'\mu}{\varepsilon-i\gamma'\mu}\left[\pi\delta(\gamma+\gamma') - iP\frac{1}{\gamma+\gamma'}\right]\\ &\quad - \frac{\varepsilon-i\gamma\mu}{\varepsilon+i\gamma\mu}\left[\pi\delta(\gamma+\gamma') + iP\frac{1}{\gamma+\gamma'}\right]\\ &\quad \left.+ \frac{\varepsilon-i\gamma\mu}{\varepsilon+i\gamma\mu}\cdot\frac{\varepsilon+i\gamma'\mu}{\varepsilon-i\gamma'\mu}\left[\pi\delta(\gamma-\gamma') + iP\frac{1}{\gamma-\gamma'}\right]\right\},\end{aligned} \tag{3.74}$$

and finally, by assuming C as real,

$$\int_0^\infty \mu U^*(x_3,\gamma)\,U(x_3,\gamma')\,dx_3 = C^2 2\pi\mu\delta(\gamma-\gamma'). \tag{3.75}$$

This relation does not hold for $\gamma = \gamma' = 0$, because the eigenfunctions vanish in this case.

The amplitude factor ${}^0C$ is obtained as

$${}^0C = \frac{2}{\mu}\sqrt{\frac{\varepsilon}{{}^0k\omega}} \tag{3.76}$$

by assuming that the surface wave mode transports unit energy and by using (3.69) and (3.28) or taking directly from (3.29) in the limit $H \to \infty$. For radiation modes with diverging energy transport we choose

$$C = \frac{1}{\sqrt{2\pi}}, \tag{3.77}$$

and the orthonormality relation is in accordance with (3.75)

$$\int_0^\infty \mu U^*(x_3, \gamma)\, U(x_3, \gamma')\, \mathrm{d}x_3 = \mu\delta(\gamma - \gamma'). \tag{3.78}$$

3.3.6. *The radiation conditions*

The expansion in terms of eigenfunctions of an open wave-guide leads to a problem connected with the so-called Sommerfeld's radiation conditions. To obtain eigenfunctions (in the generalized sense) we only required for the time being boundedness of solutions [see (3.34)]. Proceeding from the Helmholtz equation $\Delta u + k^2 u = 0$ with $k^2 = \omega^2/c^2$, it is necessary to impose additional restrictions upon the behaviour of solutions in infinity if we are interested in a class of uniquely determined solutions for an unbounded domain **B** which is the complement of a bounded one. Such restrictions are Sommerfeld's radiation conditions which are given by

$$\frac{\partial u}{\partial r} - iku = o(r^{-1/2}), \qquad u = O(r^{-1/2}) \quad \text{for} \quad r \to \infty \tag{3.79}$$

in the two-dimensional case [see e.g. Wilcox (1975)], where r is the distance from an arbitrarily fixed point and the limits are understood to be uniform with respect to all directions. Physically, the conditions imply that the solutions of the Helmholtz equation multiplied by $e^{-i\omega t}$ are also solutions of the wave equation $\Delta\Phi = \frac{1}{c^2}\ddot{\Phi}$, which behave like diverging spherical waves when $r \to \infty$. In acoustic and electro-magnetic literature, it is customary to impose the radiation conditions on the scattered field u^s only, which is expected to consist of outgoing waves and to have no sources in **B**. But of course, sources producing incoming plane waves can exist in infinity. Without modification, the scattering of waves by an obstacle within an open wave-guide cannot be achieved by this formalism, because the surface waves which appear do not satisfy the radiation conditions, especially not the second equation (3.79). After Shevchenko (1971a) the radiation condition for surface waves corresponds to the requirement for the absence of incoming waves from infinity, and that is equivalent with the modified conditions (3.79) in the onedimensional case,

$$\begin{aligned} \frac{\partial u}{\partial x_1} - iku &= o(1) \quad \text{for} \quad x_1 \to +\infty, \\ \frac{\partial u}{\partial x_1} + iku &= o(1) \quad \text{for} \quad x_1 \to -\infty, \\ u &= O(1) \quad \text{for} \quad x_1 \to \pm\infty, \end{aligned} \tag{3.80}$$

if we assume an incident wave from negative infinity. Obviously, if body and surface waves are present simultaneously, the conditions (3.79) have to be specified additionally. Furthermore, the radiation conditions in the form (3.79) (which are valid for the continuous spectrum) can only be applied to an open wave-guide which is homogeneous in the transversal direction, because only then is it possible to obtain a Helmholtz equation with a two-dimensional Laplace operator from the partial differential equation (2.17) by separation of variables. Generally an one-dimensional Helmholtz equation only can be separated, which leads to the factor $\exp\left[i(kx_1 - \omega t)\right]$. In this case, a thorough consideration of the uniqueness of solutions, together with their radiation behaviour is far beyond the scope of this book.

4. The disturbed wave-guide

4.1. General remarks

As already mentioned, the application of analytical or semi-analytical methods for the treatment of so-called irregular (disturbed) wave-guides is as difficult as it is necessary. In such wave-guides, the normal modes are reflected and scattered, and they lose their mutual independence. Disturbances occur as well in natural wave-guides, wether it be the Earth, the ocean or the atmosphere, where deviations from the "normal state" are caused by various influences, as in artificial wave-guides, e.g. in radio-technical, optical, and acoustic wave-guides. These have a beginning and an end, joints, bends, contractions of cross-section, embedded inclusions, branchings, inhomogeneities of the filling material, and roughness of the boundaries (see also Fig. 1). In all these cases irregularities occur, which must be taken into account. In the following, we will be engaged above all with discontinuities which intersperse the whole cross-section of the wave-guide, preferably with vertical discontinuities. Since surface acoustic waves at microwave frequencies have received considerable attention owing to their important applications in electronics, the discontinuity problem has become important for surface wave excitation, delay line tapping, surface wave resonators for filtering, etc. [see TUAN and CHANG (1972)].

In most active and passive microwave devices, the operation principle is based on the modification of a guided wave from a discontinuity inside a uniform wave-guide [see e.g. UZUNOGLU (1981)]. By comparing this situation with the scattering of waves at an obstacle in free space, we have to distinguish between two cases: the wave-guide with conductive walls and the dielectric wave-guide respectively. In the first case, the interaction of an incident wave with the discontinuity, compared with free-space scattering or diffraction problems, is less complicated. This is due to the fact that in the former case, the electromagnetic energy bounded in space can be represented in terms of a discrete set of eigenfunctions. On the other hand, the problem is much more complicated for dielectric wave-guides, since a discrete (surface waves) plus a continuous (radiation modes) eigenfunction set should be used with the unbounded field. We are looking at the same situation when dealing with both the closed and the open elastic wave guide (half-space).

An additional field of application is the quantitative nondesctructive testing with the help of surface waves [see e.g. MENDELSOHN et al. (1980)]. The same idea (but with greater wavelengths) has recently been applied in coal mines by using channeled in-seam seismic waves to detect faults in coal seams [see e.g. DRESEN and FREYSTÄTTER (1976)]. In a broader sense, fault detection is accomplished by teleseismic methods too (with even greater wavelengths), and precisely these are the starting-point for the considerations of the author. Some general remarks about this can be found in the introduction. A more thorough treatment will follow in section 4.5.

As already mentioned, the theory of surface waves has found broad application in recent years in the field of optical wave-guides. The parallels become immediately apparent if we replace the model "layer over half-space" by the model "half-space — layer — half-space". The surface wave modes then correspond to the guided modes and the body waves to the radiation modes. Of course, in applying optical wave-guides all activities are concentrated on minimizing the losses. The losses occur as a consequence of mode-coupling between guided modes and radiation modes, which is caused by all the various kinds of irregularities previously mentioned. Of course, in multi-mode fibres there are also conversions of one guided mode into another. On the other hand, the conversion of guided modes into radiation modes and vice versa is utilized for the input and output of light energy into and from the wave-guide respectively. To do this effectively the discontinuity problem has to be solved.

The leaking modes already mentioned also play an important part in the integrated and fibre optics [see e.g. OGUSU et al. (1980)].

In future, so-called surface polaritons and plasmons may be of interest. Surface polaritons are certain oscillations of lattice elements, which satisfy Maxwell's equations and which decay exponentially with depth. However, plasmons are related to oscillations of the elctron gas which is equivalent to the non-retarded limit case. According to MARADUDIN (1981), one can now envisage optical circuits impressed on surfaces over which surface polaritons propagate, rather similar in nature to those employed in the field of integrated optics. To assess the effectiveness of such a circuit, a study of one basic phenomenon is again required: the refraction of a surface polariton by a geometrical and/or material discontinuity.

4.2. The bra-ket formalism (interrelations with quantum theory)

4.2.1. *The ideal wave-guide*

It is well-known [see for example ALSOP (1966)] that the surface and body waves, respectively, can be compared with the eigenstates of an atom. Whereas the states of an atom are characterized by their energy, for surface waves each "state" is characterized by its phase velocity (eigen value). The ionization energy at which the continuum of the atom begins is analogous to the shear velocity of the layered half-space, which determines the cut-off of a given Love mode. In one respect the analogy breaks down. The bound states of an atom are a denumerable infinity; the surface wave modes are finite in number at any given period. Owing to these analogies it is useful to introduce a certain notation called bra-ket formalism in order to attempt to apply some methods of quantum theory to surface wave calculations. We will do so in the following, but the formalism is not maintained consistently in the course of the book. The bra-ket formalism is introduced for SH-like motion only. It has to be extended analogously to Rayleigh waves as well. Using the notations of section 2.2. let us introduce the ψ-ket vector $|\psi\rangle$, defined by

$$|\psi\rangle = \begin{pmatrix} \mathrm{i}\omega U \\ S_{12} \end{pmatrix}, \tag{4.1}$$

and the Hermitian conjugate bra vector $\langle\psi|$,

$$\langle\psi| = |\psi\rangle^{+} = (-\mathrm{i}\omega U^{*};\, S_{12}^{*}), \tag{4.2}$$

where

$$S_{12} = \mathrm{i}\mu k U(x_3) \tag{4.3}$$

is the x_3-dependent part of the 1,2-component of the stress tensor σ_{ij}. While the ψ-vector is a function of x_3 only, let the field vector Ψ be dependent on x_1, too, in such a manner that

$$\Psi = |\psi\rangle\, \mathrm{e}^{\mathrm{i}kx_1}, \tag{4.4}$$

where the time factor $\exp(-\mathrm{i}\omega t)$ is dropped as usual. By defining a matrix operator $\mathfrak{L}$ as

$$\mathfrak{L} = \begin{pmatrix} 0 & \omega/\mu \\ L_{22}/\omega & 0 \end{pmatrix}, \tag{4.5}$$

the operator equation

$$\mathfrak{L}|\psi\rangle = k|\psi\rangle \tag{4.6}$$

or

$$\mathrm{i}\frac{\partial\Psi}{\partial x_1} = -\mathfrak{L}\Psi \tag{4.7}$$

also includes the equation of motion (2.19) as the equation of definition (4.3) for S_{12}. The ψ-vector (4.1) contains an integration constant characterizing the strength of the field. To fix it conveniently a scalar product $\langle{}^{\alpha}\psi \mid {}^{\beta}\psi\rangle$ is introduced in the space of ψ-vectors

$$\langle{}^{\alpha}\psi \mid {}^{\beta}\psi\rangle = \int_0^H \langle{}^{\alpha}\psi\, |\Gamma|\, {}^{\beta}\psi\rangle\, \mathrm{d}x_3, \qquad \Gamma = \begin{pmatrix} 0 & 1 \\ 1 & 0 \end{pmatrix}, \tag{4.8}$$

i.e.,

$$\langle {}^{\alpha}\psi \mid {}^{\beta}\psi \rangle = \mathrm{i}\omega \int_0^H ({}^{\beta}U\,{}^{\alpha}S_{12}^* - {}^{\alpha}U^*\,{}^{\beta}S_{12})\,\mathrm{d}x_3$$
$$= \omega({}^{\alpha}k^* + {}^{\beta}k) \int_0^H \mu\,{}^{\alpha}U^*\,{}^{\beta}U\,\mathrm{d}x_3. \tag{4.9}$$

The operator $\mathfrak{L}$ is Hermitean with respect to this scalar product, i.e.

$$\langle \mathfrak{L}\,{}^{\alpha}\psi \mid {}^{\beta}\psi \rangle = \langle {}^{\alpha}\psi \mid \mathfrak{L}\,{}^{\beta}\psi \rangle, \tag{4.10}$$

and that means the modes belonging to different wave numbers (eigenvalues) ${}^{\alpha}k$ are mutually orthogonal. The definition (4.8) was selected in such a way that in accordance with (3.28) the scalar product between normal modes yields

$$\langle {}^{\alpha}\psi \mid {}^{\beta}\psi \rangle = 4\,{}^{\alpha}S\delta_{\alpha\beta}. \tag{4.11}$$

With regard to this definition, we have to consider some peculiarities. The scalar product of a mode with itself is identical to the energy flux on the whole, and so it can become positive and negative as well depending on the direction of propagation, i.e. an indefinite metric applies here. Further, the energy transport of modes belonging to imaginary eigenvalues is equal to zero, with the consequence that the scalar product is singular in this case (see 4.9!). For this reason imaginary eigenvalues may occur despite the hermiticity of $\mathfrak{L}$. Formally, these difficulties could be removed by introducing $\tilde{\Gamma}$,

$$\tilde{\Gamma} = \begin{pmatrix} 0 & \dfrac{{}^{\alpha}k^*}{|{}^{\alpha}k|} \\ \dfrac{{}^{\alpha}k}{|{}^{\alpha}k|} & 0 \end{pmatrix}, \tag{4.12}$$

instead of Γ, so that we would obtain a modified positively definite scalar product:

$$\langle {}^{\alpha}\psi \mid {}^{\beta}\psi \rangle = 2\omega\,|{}^{\alpha}k| \int_0^H \mu\,{}^{\alpha}U^*\,{}^{\beta}U\,\mathrm{d}x_3. \tag{4.13}$$

However, other difficulties would occur, because the eigenvalues have already been used in defining the scalar product.

For many physical problems, the system under consideration is composed of subsystems interacting with each other. In our case we have waves propagating both in positive and negative directions. Now we have to consider the interaction between these two types of waves. Often it is convenient to proceed from the spaces of independent sub-systems and then to build up the so-called product space of the composed system including the interaction between the subsystems. The two subspaces are denoted by (+) and (−) corresponding to the direction of propagation, and symbolic indices $p, q \stackrel{!}{=} \{+\ -\}$ are used, respectively. In our case, the interaction between these two subspaces is caused by the discontinuities. Note from equation (4.9) that the modes from these subspaces are mutually orthogonal, so that for normal modes the orthonormality relation (4.11) can be extended in the manner

$$\langle {}^\alpha\psi^{(p)} \mid {}^\beta\psi^{(q)} \rangle = 4\,{}^\alpha S^{(p)} \delta_{\alpha\beta} \delta_{pq} \quad \text{with} \quad \operatorname{sgn} {}^\alpha S^{(p)} = \operatorname{sgn} {}^\alpha k^{(p)}. \tag{4.14}$$

By taking into account (4.9) and (3.22) we realize that the modes with imaginary eigenvalues, i.e. $\alpha, \beta > M_s$, satisfy the relations

$$\langle {}^\alpha\psi^{(+)} \mid {}^\beta\psi^{(+)} \rangle = 0 \quad \text{and} \quad \langle {}^\alpha\psi^{(-)} \mid {}^\beta\psi^{(-)} \rangle = 0, \tag{4.15}$$

as well as by using the abbreviation

$$ {}^\alpha s = \frac{\omega}{2}\, |{}^\alpha k| \int_0^H \mu\, {}^\alpha U^* {}^\alpha U \, \mathrm{d}x_3, \quad {}^\alpha s = |{}^\alpha S| \quad \text{for} \quad \alpha < M_s, \tag{4.16}$$

$$\langle {}^\alpha\psi^{(+)} \mid {}^\beta\psi^{(-)} \rangle = -4\mathrm{i}\delta_{\alpha\beta}\, {}^\alpha s \quad \text{and} \quad \langle {}^\alpha\psi^{(-)} \mid {}^\beta\psi^{(+)} \rangle = 4\mathrm{i}\delta_{\alpha\beta}\, {}^\alpha s. \tag{4.17}$$

Summarizing (4.14), (4.15), and (4.17) the orthonormality relation may be written as

$$\langle {}^\alpha\psi^{(p)} \mid {}^\beta\psi^{(q)} \rangle = 4\delta_{\alpha\beta}\, {}^\alpha s \operatorname{sgn} q [\delta_{pq} H(M_s - \alpha) + \mathrm{i}(1 - \delta_{pq})\, H(\alpha - M_s)], \tag{4.18}$$

where Heaviside's function $H(x)$ is used, defined by

$$H(x) = \begin{cases} 1 & \text{for} \quad x \geqq 0 \\ 0 & \text{for} \quad x < 0' \end{cases} \tag{4.19}$$

and it holds that $\operatorname{sgn} q = +1$ for $q = (+)$ and $\operatorname{sgn} q = -1$ for $q = (-)$. Up to now we have used the tacit understanding that the wave-guide is closed. Dropping this assumption we have to consider "continuous eigenvectors" $|\psi(\gamma)\rangle$ as well. Analogous to (4.9), we obtain

$$\begin{aligned} \langle \psi(\gamma) \mid \psi(\gamma') \rangle &= \mathrm{i}\omega \int_0^\infty [U(x_3, \gamma')\, S_{12}^*(x_3, \gamma) \\ &\quad - U^*(x_3, \gamma)\, S_{12}(x_3, \gamma')]\, \mathrm{d}x_3 \\ &= \omega(k' + k^*) \int_0^\infty \mu U(x_3, \gamma')\, U^*(x_3, \gamma)\, \mathrm{d}x_3, \end{aligned} \tag{4.20}$$

where for the present γ is defined more generally as in (3.4) by

$$\gamma^2 = \Lambda^2 - k^2. \tag{4.21}$$

Note that k may be real or imaginary, and by assuming

$$\frac{\omega}{2}|k|\int_0^\infty \mu U(x_3,\gamma')\,U^*(x_3,\gamma)\,\mathrm{d}x_3 = S(\gamma)\,\delta(\gamma-\gamma'),\qquad S(\gamma)>0 \tag{4.22}$$

it follows [compare with (4.18)!] that

$$\begin{aligned}\langle\psi^{(p)}(\gamma)\mid\psi^{(q)}(\gamma')\rangle &= 4S(\gamma)\,\mathrm{sgn}\,q\delta(\gamma-\gamma')\\ &\quad\times[\delta_{pq}H(\Lambda-\gamma)+\mathrm{i}(1-\delta_{pq})\,H(\gamma-\Lambda)].\end{aligned} \tag{4.23}$$

Additionally, the mixed scalar products between discrete and continuous eigenvectors vanish, i.e.

$$\langle\psi^{(p)}(\gamma)\mid{}^{\alpha}\psi^{(q)}\rangle = 0 \quad\text{and}\quad \langle{}^{\alpha}\psi^{(p)}\mid\psi^{(q)}(\gamma')\rangle = 0. \tag{4.24}$$

4.2.2. *The wave-guide with weak disturbances*

4.2.2.1. Dirac's perturbation theory

To study wave propagation in weakly disturbed wave-guides, it is possible to apply a method similar to the one called variation of constants for solution of linear differential equations. Such a perturbation theory was introduced into quantum theory by Dirac 1926 to consider explicitly time-dependent perturbations. This theory can also be applied for our purposes by replacing the time-dependent perturbations by x_1-dependent ones. The perturbations in x_1-direction can be imagined as changes of the elastic parameters, the cross-section, or the impedance along the wave-guide. The perturbations are assumed to allow a splitting of the operator $\mathfrak{L}$ into an undisturbed part $\mathfrak{L}^0$ (being valid for the ideal wave-guide) and into the perturbation operator $\mathfrak{L}^s$:

$$\mathfrak{L} = \mathfrak{L}^0 + \mathfrak{L}^s. \tag{4.25}$$

In the next section, we will describe how to carry out this splitting in a practical manner. Here, it is only assumed that $\mathfrak{L}^s$ contains no derivatives by x_1.

As already mentioned, mode coupling is an important effect in disturbed wave-guides. According to Marcuse (1974) there are many different ways of obtaining coupled-mode equations. We will mention only two of these. The first method will be discussed in some detail. It is based on the assumption that the field Ψ^s within the disturbed wave-guide may be expanded in terms of modes ${}^{\nu}\Psi^0$ of the undisturbed wave-guide. By using the space-dependent expansion coefficients $C_\nu(x_1)$, let the expansion be

$$\Psi^s = \sum_\nu\!\!\!\!\!\!\int C_\nu(x_1)\,{}^{\nu}\Psi^0, \tag{4.26}$$

where the symbol $\mathop{\mathcal{S}}$ indicates that it has to be summed up over the discrete part of the spectrum and to be integrated over the continuous one, while taking into account forward and backward travelling waves. Substituting Ψ^s into the equation of motion (4.7) it follows that

$$\mathrm{i}\,\frac{\partial}{\partial x_1}\mathop{\mathcal{S}}_{\nu} C_\nu(x_1)\,{}^\nu\Psi^0 = -\mathfrak{L}^0\mathop{\mathcal{S}}_{\nu} C_\nu(x_1)\,{}^\nu\Psi^0 - \mathfrak{L}^s\mathop{\mathcal{S}}_{\nu} C_\nu(x_1)\,{}^\nu\Psi^0, \tag{4.27}$$

and by assuming commutativity of summation (or integration) and differentiation we obtain

$$\mathrm{i}\mathop{\mathcal{S}}_{\nu}\left(\frac{\mathrm{d}C_\nu}{\mathrm{d}x_1}\,{}^\nu\Psi^0 + C_\nu\,\frac{\partial^\nu\Psi^0}{\partial x_1}\right) = -\mathop{\mathcal{S}}_{\nu} C_\nu(x_1)\,\mathfrak{L}^{0\nu}\Psi^0 - \mathop{\mathcal{S}}_{\nu} C_\nu(x_1)\,\mathfrak{L}^{s\nu}\Psi^0. \tag{4.28}$$

Taking into account that the eigenfunctions of the undisturbed wave-guide satisfy the equation of motion with the undisturbed operator, (4.28) is simplified in the manner:

$$\mathrm{i}\mathop{\mathcal{S}}_{\nu}\frac{\mathrm{d}C_\nu}{\mathrm{d}x_1}\,{}^\nu\Psi^0 = -\mathop{\mathcal{S}}_{\nu} C_\nu(x_1)\,\mathfrak{L}^{s\nu}\Psi^0. \tag{4.29}$$

Introducing the eigenvectors ${}^\nu\psi^0$ depending on x_3 only [see (4.4)] and forming the scalar product with $\langle{}^\mu\psi^0|$, it follows that

$$\mathrm{i}\mathop{\mathcal{S}}_{\nu}\frac{\mathrm{d}C_\nu}{\mathrm{d}x_1}\langle{}^\mu\psi^0\mid{}^\nu\psi^0\rangle\,\mathrm{e}^{\mathrm{i}^\nu kx_1} = -\mathop{\mathcal{S}}_{\nu} C_\nu(x_1)\,\langle{}^\mu\psi^0\mid\mathfrak{L}^{s\nu}\psi^0\rangle\,\mathrm{e}^{\mathrm{i}^\nu kx_1}. \tag{4.30}$$

By using the relation

$$\langle{}^\mu\psi^0\mid{}^\nu\psi^0\rangle = P(\mu,\nu)\,\delta(\mu,\nu), \tag{4.31}$$

which summarizes the orthonormality relations (4.18), (4.23), and (4.24), we obtain

$$\mathrm{i}\mathop{\mathcal{S}}_{\nu}\frac{\mathrm{d}C_\nu}{\mathrm{d}x_1}P(\mu,\nu)\,\delta(\mu,\nu)\,\mathrm{e}^{\mathrm{i}^\nu kx_1} = -\mathop{\mathcal{S}}_{\nu} C_\nu(x_1)\,\langle{}^\mu\psi^0\mid\mathfrak{L}^{s\nu}\psi^0\rangle\,\mathrm{e}^{\mathrm{i}^\nu kx_1}, \tag{4.32}$$

where $\delta(\mu,\nu)$ has to be interpreted as Kronecker's symbol for discrete modes, and as Dirac's δ-function for continuous ones. The quantity $P(\mu,\nu)$ includes the complex character of all orthonormality relations, and it has to be specified from case to case. Defining matrix elements $M_{\mu\nu}$ in the manner

$$M_{\mu\nu} = \frac{\mathrm{i}}{P(\mu,\mu)}\langle{}^\mu\psi^0\mid\mathfrak{L}^{s\nu}\psi^0\rangle, \tag{4.33}$$

equation (4.32) yields the following system of differential equations for the unknown expansion coefficients $C_\nu(x_1)$ depending on x_1:

$$\frac{\mathrm{d}C_\mu}{\mathrm{d}x_1} = \sum_\nu M_{\mu\nu} C_\nu \, \mathrm{e}^{\mathrm{i}({}^\nu k - {}^\mu k) x_1}. \tag{4.34}$$

Up to this point the theory is correct. The approximation begins now, because it is impossible to solve this infinite linear system of equations rigorously. Let us expand the perturbation operator $\mathfrak{L}^s$ as well as the coefficients C_ν in terms of a small perturbation parameter s,

$$\mathfrak{L}^s = s\mathfrak{L}^{s(1)} + s^2\mathfrak{L}^{s(2)} + \cdots, \tag{4.35}$$

$$C_\nu = C_\nu^{(0)} + sC_\nu^{(1)} + s^2C_\nu^{(2)} + \cdots, \tag{4.36}$$

and substitute into (4.34). By using the modified matrix elements $M_{\mu\nu}^{(j)}$, defined by

$$M_{\mu\nu}^{(j)} = \frac{\mathrm{i}}{P(\mu, \nu)} \langle {}^\mu\psi^0 \mid \mathfrak{L}^{s(j)\,\nu}\psi^0 \rangle, \quad j = 1, 2, \ldots, \tag{4.37}$$

the method of coefficient comparison yields the following equations for the coefficients from 0th up to 2nd order:

$$\frac{dC_\mu^{(0)}}{\mathrm{d}x_1} = 0, \tag{4.38}$$

$$\frac{\mathrm{d}C_\mu^{(1)}}{\mathrm{d}x_1} = \sum_\nu C_\nu^{(0)} M_{\mu\nu}^{(1)} \, \mathrm{e}^{\mathrm{i}({}^\nu k - {}^\mu k) x_1}, \tag{4.39}$$

$$\frac{\mathrm{d}C_\mu^{(2)}}{\mathrm{d}x_1} = \sum_\nu \{C_\nu^{(1)} M_{\mu\nu}^{(1)} + C_\nu^{(0)} M_{\mu\nu}^{(2)}\} \, \mathrm{e}^{\mathrm{i}({}^\nu k - {}^\mu k) x_1}, \tag{4.40}$$

which can be solved by integrations in a simple manner, provided that there are only constants on the right sides with the exception of the exponential factors. Generally, the first approximation already yields a satisfying accuracy, if the effect of $\mathfrak{L}^s$ is not too great. In accordance with (4.38), the expansion coefficients can be treated as weakly-varying quantities for small perturbations, and that also follows by setting $M_{\mu\nu}$ identically equal to zero (undisturbed case). If the quantities C_ν are given by initial conditions for $x_1 = x_0$, the $C_\nu(x_1)$ can also be found using the method of successive approximation, i.e.

$$C_\mu^{(j+1)}(x_1) = C_\mu(x_0) + \int_{x_0}^{x_1} \sum_\nu M_{\mu\nu} C_\nu^{(j)}(x) \, \mathrm{e}^{\mathrm{i}({}^\nu k - {}^\mu k) x} \, \mathrm{d}x, \quad j = 0, 1, 2, \ldots \tag{4.41}$$

In quantum theory, the most important observable is the transition probability of a given physical system, which changes from a prescribed time-independent initial state into another state within a given time interval. These transition propabilities correspond to the coefficients of mode coupling in our theory. Applying an appropriate normalization the latter ones are equal to the square of the expansion coefficients. In this connection we have to refer to the theory of the so-called S-matrix (scattering matrix), which is in a certain sense an extension of Dirac's perturbation theory. The scattering matrix is an unitary operator S, which transforms the totality of all possible initial states $|j\rangle$ into

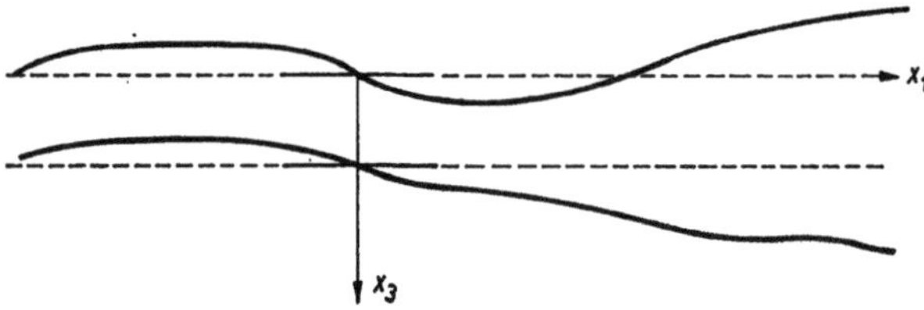

Fig. 17. Expansion in terms of local normal modes

the totality of final states $|f\rangle$ generated by a prescribed interaction. This transformation or scattering is described by the matrix elements $S_{fj} = \langle f |S| j\rangle$. The quantity $w_{fj} = |S_{fj}|^2$ is the transition probability of the process $|j\rangle \to |f\rangle$ which is proportional to the cross-section. On the whole, S_{fj} is identical with the scattering amplitude of the considered process. A similar theory can also be developed for disturbed wave-guides.

For the derivation of coupled-mode equations, another important possibility is based on the application of so-called local modes instead of modes of the ideal wave-guide. Let us consider a wave-guide with an x_1-dependent cross-section. We can now express the field at each point x_1 in terms of modes of a hypothetical wave-guide, whose width coincides locally with the width of the actual wave-guide (see Fig. 17).

These local modes themselves are not solutions of the equations of motion, since their parameters are functions of x_1. However, they can be superimposed to yield a solution to the equations of motion that represents the field of the actual wave-guide that we want to describe. Proceeding along these lines it is possible to derive coupled-mode equations which are similar to (4.34). MARCUSE (1974) discussed the differences of the coupling coefficients for the two methods in great detail by considering some special cases. Both types of mode expansions have their advantages for certain types of problems. The ideal mode expansion is particularly well suited to wave-guides with perfect geometry but with inhomogeneities of material parameters. The expansion into local normal modes is useful for the description of wave-guides with imperfect geometry.

4.2.2.2. Applications of the perturbation theory

4.2.2.2.1. General remarks

The quantum theoretical perturbation theory has found many applications in the theory of wave-guides. Approximative solutions of coupled-mode equations are discussed in the above-mentioned book of Marcuse (1974) for various special dielectric wave-guides. The so-called cross-sectional method, well-known from the Soviet literature (metod poperechnych sechenii), is based on the expansion in terms of local normal modes. The monographs of Katsenelenbaum (1961) and Shevchenko (1971a) contain fundamental contributions to this set of problems for closed and open wave-guides as well. A mathematically rigorous foundation of the cross-sectional method was obtained by Fedorjuk (1973) for inhomogeneous acoustic wave-guides, and Abram (1974), who applied the method of coupled modes to elastic wave-guides considering the local modes as quasimodes, which are "almost" eigenmodes of the real guide for slow variations of the wave-guide parameters.

For the first time, Kennett (1984) applied the perturbation theory of coupled modes to the propagation of seismic surface waves in a laterally varying crust (see section 4.5.3.2.6.). These few references do not nearly exhaust the possibilities of application of quantum mechanical perturbation theory in wave-guide theory, but they have to be sufficient here.

4.2.2.2.2. Separation of the perturbation operator $\mathfrak{L}^s$ for disturbances of material parameters

Let us understand by disturbances of material parameters the x_1-dependences of the quantities μ and ϱ. The equation of motion generalized in this respect is

$$\frac{\partial}{\partial x_1}\left[\mu(x_1, x_3)\,\frac{\partial u_2}{\partial x_1}\right] + \frac{\partial}{\partial x_3}\left[\mu(x_1, x_3)\,\frac{\partial u_2}{\partial x_3}\right] = \varrho(x_1, x_3)\,\ddot{u}_2. \tag{4.42}$$

To apply the perturbation method, it is necessary that μ and ϱ be expressed in the manner

$$\mu = \mu^0(x_3) + \mu^s(x_1, x_3), \qquad \varrho = \varrho^0(x_3) + \varrho^s(x_1, x_3) \tag{4.43}$$

with

$$|\mu^s/\mu^0| \ll 1 \quad \text{and} \quad |\varrho^s/\varrho^0| \ll 1. \tag{4.44}$$

While the operator component L_{22} can be separated immediately into an undisturbed part and a perturbation one, the quantity ω/μ has to be expanded in a series, which is terminated after the second term due to (4.44):

$$\frac{\omega}{\mu} = \frac{\omega}{\mu_0} - \omega\left(\frac{\mu^s}{\mu_0^2} + \cdots\right). \tag{4.45}$$

Accordingly, for $\mathfrak{L}$ the representation

$$\mathfrak{L} = \mathfrak{L}^0 + \mathfrak{L}^s = \begin{pmatrix} 0 & \dfrac{\omega}{\mu^0} \\ \dfrac{L_{22}^0}{\omega} & 0 \end{pmatrix} + \begin{pmatrix} 0 & -\omega \dfrac{\mu^s}{\mu_0^2} \\ \dfrac{L_{22}^s}{\omega} & 0 \end{pmatrix} \tag{4.46}$$

is obtained with

$$\begin{aligned} L_{22}^0 &= \frac{\partial}{\partial x_3}\left(\mu^0 \frac{\partial}{\partial x_3}\right) + \varrho^0\omega^2 \quad \text{and} \\ L_{22}^s &= \frac{\partial}{\partial x^3}\left(\mu^s \frac{\partial}{\partial x_3}\right) + \varrho^s\omega^2 . \end{aligned} \tag{4.47}$$

4.2.2.2.3. The "transient function"

To maintain the language of quantum theory let us now discuss the so-called "transient function", which is understood in our case as the rate of change of material parameters in direction x_1. We rely closely on the representation of Landau and Lifschitz (1965). Let the system be in the n-th eigen state before the disturbance, i.e. mode with number n meets at the disturbed area from the left, and it follows

$$C_{\nu n}^{(0)} = \delta_{\nu n} . \tag{4.48}$$

For the present, let the disturbance be confined to a bounded x_1-interval, then the system is in the stationary state ${}^n\Psi^{(0)}$ for $x_1 \to -\infty$. For any x_1 the system is described by the function $\Psi^s = \sum_\nu C_{\nu n}(x_1)\, {}^\nu\Psi^{(0)}$, and according to (4.41) the first approximation of $C_{\nu n}$ is obtained as

$$C_{\nu n}(x_1) \approx C_{\nu n}^{(1)}(x_1) = \delta_{\nu n} + \int_{-\infty}^{x_1} M_{\nu n} \quad e^{i\omega_{n\nu}x}\, dx , \tag{4.49}$$

with

$$\omega_{n\nu} = {}^n k - {}^\nu k . \tag{4.50}$$

After the action of the disturbance (i.e. for $x_1 \to \infty$) the coefficients $C_{\nu n}$ take the constant values $C_{\nu n}(\infty)$, and the system is in a state with the wave function $\Psi_\infty = \sum_\nu C_{\nu n}(\infty)\, {}^\nu\Psi^{(0)}$ again satisfying the undisturbed equation

$$i \frac{\partial \Psi_\infty}{\partial x_1} = -\mathfrak{L}^0 \Psi_\infty , \tag{4.51}$$

but being different from the original function. In quantum theory, the algebraic norm of $C_{\nu n}(\infty)$ is equal to the probability that the system is in the n-th stationary state. In wave-guide physics this quantity describes the fact that under the influence of the disturbance, the incoming n-th mode is partly transformed into the ν-th mode. That is to say, the transformation coefficient of the n-th mode into the ν-th one is given by

$$w_{\nu n} = \left| \int_{-\infty}^{\infty} M_{\nu n} \, \mathrm{e}^{\mathrm{i}\omega_{n\nu}x} \, \mathrm{d}x \right|^2 . \tag{4.52}$$

Now, let us consider a disturbance which, after its occurrence, continues for ever, i.e. for $x_1 \to -\infty$ it follows that $\mathfrak{L}^s \to 0$, and for $x_1 \to \infty$ $\mathfrak{L}^s$ tends to a finite limit different from zero. In this case, the integral in (4.52) diverges. This divergence, however, is physically unimportant, and it can be removed by partial integration as follows:

$$\int_{-\infty}^{x_1} M_{\nu n} \mathrm{e}^{\mathrm{i}\omega_{n\nu}x} \, \mathrm{d}x = \left[M_{\nu n} \frac{\mathrm{e}^{\mathrm{i}\omega_{n\nu}x}}{\mathrm{i}\omega_{n\nu}} \right]_{-\infty}^{x_1} - \int_{-\infty}^{x_1} \frac{\partial M_{\nu n}}{\partial x_1} \cdot \frac{\mathrm{e}^{\mathrm{i}\omega_{n\nu}x}}{\mathrm{i}\omega_{n\nu}} \, \mathrm{d}x . \tag{4.53}$$

The first term on the right side vanishes at the lower limit and yields a corection of the original eigenfunction ${}^n\Psi^{(0)}$ for $x_1 \to \infty$ under the influence of the constant part of the disturbance which has no influence on the mode conversion. The transformation coefficient, however, is given by the square of the second term:

$$w_{\nu n} = \frac{1}{\omega_{n\nu}^2} \left| \int_{-\infty}^{\infty} \frac{\partial M_{\nu n}}{\partial x_1} \, \mathrm{e}^{\mathrm{i}\omega_{n\nu}x} \, \mathrm{d}x \right|^2 . \tag{4.54}$$

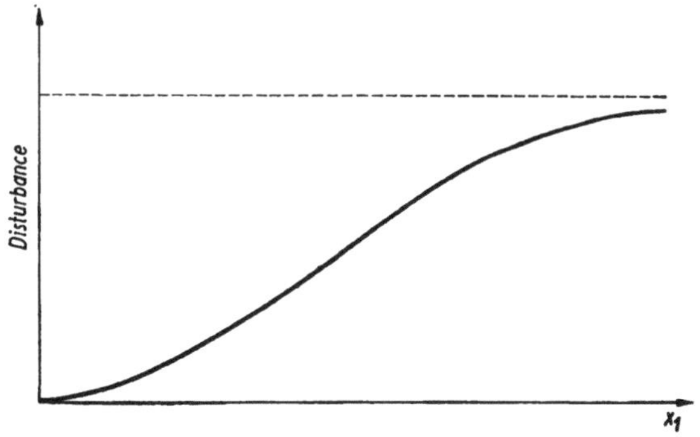

Fig. 18. "Adiabatic" transient function

The further evaluation of the integral in (4.54) depends decisively on the "transient function", i.e. on the character of the functions μ^s and ϱ^s. We can distinguish two limit cases:

a) Slow "switching on" of the disturbance (Fig. 18).
If there is only a small change of disturbance within a x_1-interval of the order $1/\omega_{n\nu}$, the integration in (4.54) can be carried out as follows:

$$w_{\nu n}^{(\mathrm{ad})} = \frac{1}{\omega_{n\nu}^2}\left|\lim_{L\to\infty}\int_{-L}^{L}\frac{\partial M_{\nu n}}{\partial x_1}\,\mathrm{e}^{\mathrm{i}\omega_{n\nu}x}\,\mathrm{d}x\right|^2$$

$$= \frac{1}{\omega_{n\nu}^2}\left|\frac{\partial M_{\nu n}}{\partial x_1}\right|^2\cdot\left|\lim_{L\to\infty}\int_{-L}^{L}\mathrm{e}^{\mathrm{i}\omega_{n\nu}x}\,\mathrm{d}x\right|^2$$

$$= \frac{4}{\omega_{n\nu}^2}\left|\frac{\partial M_{\nu n}}{\partial x_1}\right|^2\cdot\left|\lim_{L\to\infty}\frac{\sin\omega_{n\nu}L}{\omega_{n\nu}}\right|^2$$

$$\overset{!}{=} \frac{4\pi^2}{\omega_{n\nu}^2}\left|\frac{\partial M_{\nu n}}{\partial x_1}\right|^2\cdot\delta^2(\omega_{n\nu}). \tag{4.55}$$

Thus, if we assume additionally $\omega_{n\nu} \neq 0$, i.e. there is no degeneracy, we obtain $w_{\nu n}^{(\mathrm{ad})} \ll 1$ under these circumstances. That is to say, if the disturbance is "switched on" adiabatically, a mode of the undisturbed wave-guide changes gradually into the corresponding one of the disturbed wave-guide.

b) Abrupt "switching on" of the disturbance (Fig. 19).
If the disturbance is "switched on" abruptly, that is to say, within an x_1-interval which is small in comparison with $1/\omega_{n\nu}$ (e.g. a vertical discontinuity), the derivatives $\dfrac{\partial M_{\nu n}}{\partial x_1}$ tend to infinity, and it is possible to take the slowly

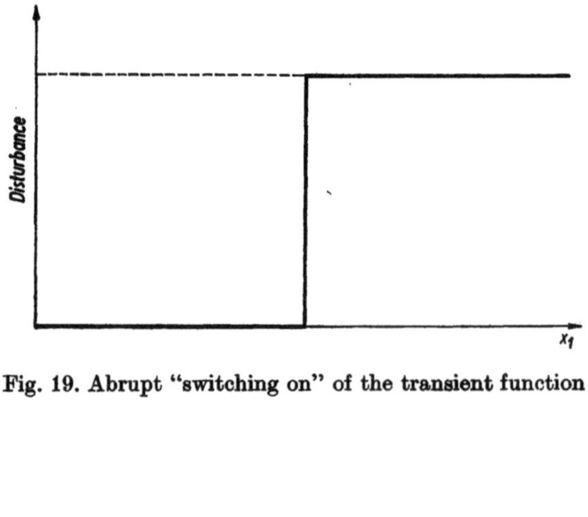

Fig. 19. Abrupt "switching on" of the transient function

varying factor exp $(i\omega_{n\nu}x)$ before the integral in (4.54), and we obtain

$$w_{\nu n}^{(\mathrm{sud})} = \frac{|M_{\nu n}|^2}{\omega_{n\nu}^2}. \tag{4.56}$$

Using this formula, the mode-transformation coefficients may be determined for small abrupt disturbances. But in reality, there are also comparatively abrupt great changes of material parameters. It is still possible to determine the transformation coefficients in this case too, in case the undisturbed eigenfunctions ${}^{\nu}\psi^0$ cannot adapt so quickly to the abrupt change. Consequently, they continue a certain distance after the material jump, but no longer as eigenfunctions of the disturbed operator $\mathfrak{L}$. Then it is possible to expand the undisturbed eigenfunctions in terms of eigenfunctions of $\mathfrak{L}$, and the algebraic norms of the expansion coefficients are the transformation coefficients sought for. It can be shown, that the formula obtained in this way turns into (4.56) for small disturbances. These brief remarks has to be sufficient here.

Let us now modify the approach by asking for the effect of a spatially constant disturbance as a function of the distance passed through, rather than for the influence of the transient process. Let the matrix element $M_{\nu n}$ be constant in the x_1-interval $[0, L]$. Then by using

$$D(\omega_{n\nu}, L) = \frac{1 - \cos \omega_{n\nu} L}{\omega_{n\nu}^2} = \frac{\sin^2 \dfrac{\omega_{n\nu} L}{2}}{\dfrac{\omega_{n\nu}^2}{2}}, \tag{4.57}$$

it follows from (4.52) that

$$w_{\nu n} = 2M_{\nu n}^2 D(\omega_{n\nu}, L). \tag{4.58}$$

Due to

$$\frac{1}{\pi} \lim_{L\to\infty} \frac{\sin^2 Lx}{Lx^2} = \delta(x), \tag{4.59}$$

this equation can also be written as

$$w_{\nu n} = 2\pi M_{\nu n}^2 L\, \delta(\omega_{n\nu}), \tag{4.60}$$

i.e., $w_{\nu n}$ grows linearly with the distance L passed through; the wave-theoretical analogy of Fermi's "golden rule" expressing the fact that the conditional probability increases linearly with time for a temporally constant disturbance.

With certain modifications, the formulas of this section are also valid for transitions from the discrete into the continuous spectrum and vice versa, and for transitions within the continuous spectrum.

4.2.2.2.4. Application of the perturbation theory to a discontinuity of surface impedance

If the disturbance is caused by an x_1-dependence of the impedance parameter ε, i.e. $\varepsilon = \varepsilon(x_1)$, then the perturbation theory, which is based on the eigenfunctions of the ideal wave-guide, is not easily applicable, because the boundary conditions are disturbed in this case and therefore no perturbation operator $\mathfrak{L}^s$ can be separated. On the other hand, in a more general sense the operator $\mathfrak{L}$ is also disturbed, because it is "generated" by a differential expression *and* by corresponding boundary conditions as well. Therefore, in one way or another Dirac's perturbation theory should also be applicable to this case. However, the expansion in terms of local modes already mentioned is probably more suited to this problem. The local modes $|{}^\nu\varphi\rangle$ depending now on x_1 too, satisfy the equation of motion

$$\mathfrak{L}^0 \, |{}^\nu\varphi\rangle = {}^\nu k(x_1) \, |{}^\nu\varphi\rangle \tag{4.61}$$

as well as the simplified written orthonormality relation

$$\langle{}^\mu\varphi \mid {}^\nu\varphi\rangle = \bar{P}(\mu, \nu) \, \delta(\mu, \nu) \tag{4.62}$$

for every cross-sectional area (i.e. $x_1 = \text{const.}$). Note that the eigenvalues in (4.61) are also space-dependent owing to $\varepsilon = \varepsilon(x_1)$. The solution for the disturbed wave-guide is built up by the local modes as follows:

$$\Psi^s = \sum_\nu\!\!\!\!\!\!\int D_\nu(x_1) \, \mathrm{e}^{\mathrm{i}\int^{x_1} {}^\nu k(x)\mathrm{d}x} \, |{}^\nu\varphi\rangle. \tag{4.63}$$

In the exponent, the integral occurs because of the x_1-dependence of ${}^\nu k$. The local modes do not satisfy the equation of motion

$$\mathrm{i}\frac{\partial \Psi}{\partial x_1} = -\mathfrak{L}^0 \Psi. \tag{4.64}$$

That is why the space-dependent expansion coefficients D_ν have to be chosen so that Ψ^s satisfies this equation. Substituting the statement (4.63) into the equation of motion (4.64) and considering (4.61) we obtain

$$\mathrm{i} \sum_\nu\!\!\!\!\!\!\int \frac{\mathrm{d}D_\nu}{\mathrm{d}x_1} \, |{}^\nu\varphi\rangle \, \mathrm{e}^{\mathrm{i}\int^{x_1} {}^\nu k \mathrm{d}x} = -\mathrm{i} \sum_\nu\!\!\!\!\!\!\int D_\nu \, \mathrm{e}^{\mathrm{i}\int^{x_1} {}^\nu k \mathrm{d}x} \, \frac{\partial \, |{}^\nu\varphi\rangle}{\partial x_1}. \tag{4.65}$$

By taking the scalar product with $\langle{}^\mu\varphi|$ and considering (4.62) a system of integro-differential equations

$$\frac{\mathrm{d}D_\mu}{\mathrm{d}x_1} \, \bar{P}(\mu, \mu) = - \sum_\nu\!\!\!\!\!\!\int D_\nu \, \mathrm{e}^{\mathrm{i}\int^{x_1} ({}^\nu k - {}^\mu k)\mathrm{d}x} \, \langle{}^\mu\varphi| \, \frac{\partial \, |{}^\nu\varphi\rangle}{\partial x_1} \tag{4.66}$$

follows for the determination of D_ν. For an incident surface wave, SHEVCHENKO (1971a) obtained such a system for the coefficients sought for in a similar manner. By series expansion he succeeded in finding solutions for this system for a slowly varying impedance and for a small impedance jump, respectively. A purely numerical evaluation of the system yielded statements for a quick varying impedance or a great impedance jump as well. We shall not go into details here, because the same problem will be treated later by applying other methods.

MARCUSE (1972) realized that for certain problems (e.g. a precipitous change of the wave-guide diameter) neither the expansion in terms of modes of the ideal wave-guide nor in terms of local modes is convenient. In such cases the mode-matching technique is more favourable. We will represent its application for the impedance-jump problem in great detail.

4.3. The closed wave-guide with a vertical discontinuity or an impedance jump, respectively

4.3.1. *General remarks*

The closed wave-guide (type I or II) is considered to be an exceptionally simple and clear model capable for guiding surface wave-like modes according to section 3.2., but there are no continuous radiation modes. Incidently, Maradudin [see STEGEMAN et al. (1981)] surpressed the latter ones by introducing a metallized boundary in the vacuum for the problem of the scattering of surface polaritons. That means it was explicitely attempted to change from the open into the closed wave-guide in order to obtain numerically more stable results. The results suggest that for this special problem the reflection and transmission coefficients are not very sensitive to the use of a finite value of the space between the metallized boundary and the dielectric. Consequently, only a slight falsification of the results need be expected in this case.

In our case, the wave-guide is characterized by the parameters μ, ϱ, H, and ε. Let it consist of the regions (1) and (2) welded together at the boundary $x_1 = 0$, where at least one of the parameters mentioned above undergoes a jump. If it is the parameter H, we have a wave-guide with a surface step (Fig. 1a), and a vertical discontinuity for changing μ and ϱ (Fig. 1b). If the impedance parameter ε alone is changing, we are dealing with the impedance jump model.

4.3.2. *Expansion in terms of eigenfunctions in the regions right and left of the discontinuity*

Let us suppose a slow mode is travelling from left [region (1)] to the discontinuity. Using the mode-matching procedure let us now consider the reflection and refraction of this mode by expanding the resultant wave-field right and left from the discontinuity in terms of eigenfunctions of the closed wave-guide (see section 3.2.3.). At this moment, there is no need to specify the type of the wave-guide. By using the notations of sections 2.1.2. and 3.2. and by considering formula (4.3) the resultant displacements and stresses for regions (1) and (2) are obtained as

$$u_{\text{res}}^{(1)} = {}^0U^{(1)}(x_3)\, e^{i\,{}^0k^{(1)}x_1} + \sum_{\nu=0}^{\infty} a_\nu {}^\nu U^{(1)}(x_3)\, e^{-i\,{}^\nu k^{(1)}x_1}, \tag{4.67}$$

$$u_{\text{res}}^{(2)} = \sum_{\nu=0}^{\infty} b_\nu {}^\nu U^{(2)}(x_3)\, e^{i\,{}^\nu k^{(2)}x_1}, \tag{4.68}$$

$$\sigma_{12\text{res}}^{(1)} = \mu_1 i\,{}^0k^{(1)}\,{}^0U^{(1)}(x_3)\, e^{i\,{}^0k^{(1)}x_1} - i\mu_1 \sum_{\nu=0}^{\infty} a_\nu {}^\nu k^{(1)}\,{}^\nu U^{(1)}(x_3)\, e^{-i\,{}^\nu k^{(1)}x_1}, \tag{4.69}$$

$$\sigma_{12\text{res}}^{(2)} = i\mu_2 \sum_{\nu=0}^{\infty} b_\nu {}^\nu k^{(2)}\,{}^\nu U^{(2)}(x_3)\, e^{i\,{}^\nu k^{(2)}x_1}. \tag{4.70}$$

The expansion coefficients a_ν and b_ν of regions (1) and (2) are respectively, generalized reflection and transmission coefficients. They refer to the incident slow mode, and they are generalized in regard to the usual meaning, because they depend yet on the normalization used here. The normalization is carried out according to section 3.2.4. The normalization constants have been temporarily incorporated into the functions ${}^\nu U^{(1)}$ and ${}^\nu U^{(2)}$ ($\nu = 0, 1, 2, \ldots$). They will be considered only in calculating the corresponding scalar products.

4.3.3. *Setting up an infinite system of equations for the reflection and transmission coefficients of the slow mode for wave-guide type I*

To eliminate the coefficients a_ν and b_ν, we proceed from the continuity conditions on the plane of discontinuity $x_1 = 0$,

$$u_{\text{res}}^{(1)} = u_{\text{res}}^{(2)} \quad \text{and} \quad \sigma_{12\text{res}}^{(1)} = \sigma_{12\text{res}}^{(2)} \quad \text{for} \quad x_1 = 0, \tag{4.71}$$

i.e. by using (4.67) —(4.70)

$${}^0U^{(1)}(x_3) + \sum_{\nu=0}^{\infty} a_\nu {}^\nu U^{(1)}(x_3) = \sum_{\nu=0}^{\infty} b_\nu {}^\nu U^{(2)}(x_3), \tag{4.72}$$

$${}^0k^{(1)}\,{}^0U^{(1)}(x_3) - \sum_{\nu=0}^{\infty} a_\nu {}^\nu k^{(1)}\,{}^\nu U^{(1)}(x_3) = \frac{\mu_2}{\mu_1} \sum_{\nu=0}^{\infty} b_\nu {}^\nu k^{(2)}\,{}^\nu U^{(2)}(x_3). \tag{4.73}$$

The normalization is different for slow and fast modes according to (3.23) and (3.24). Therefore it is convenient to split the sums as follows:

$$ {}^0U^{(1)} + a_0 {}^0U^{(1)} + \sum_{\nu=1}^{\infty} a_\nu {}^\nu U^{(1)} = b_0 {}^0U^{(2)} + \sum_{\nu=1}^{\infty} b_\nu {}^\nu U^{(2)}, \tag{4.74} $$

$$ {}^0k^{(1)0}U^{(1)} - a_0 {}^0k^{(1)0}U^{(1)} - \sum_{\nu=1}^{\infty} a_\nu {}^\nu k^{(1)\nu} U^{(1)} = \frac{\mu_2}{\mu_1} b_0 {}^0k^{(2)0}U^{(2)} + \frac{\mu_2}{\mu_1} \sum_{\nu=1}^{\infty} b_\nu {}^\nu k^{(2)\nu} U^{(2)}. \tag{4.75} $$

Further, the "mixed scalar products" between the eigenfunctions of regions (1) and (2) are introduced in the manner

$$ P = \int_0^H {}^0U^{(1)0}U^{(2)*}\,\mathrm{d}x_3, \qquad Q_\nu = \int_0^H {}^0U^{(2)*\nu}U^{(1)}\,\mathrm{d}x_3, $$

$$ R_\nu = \int_0^H {}^0U^{(1)\nu}U^{(2)*}\,\mathrm{d}x_3, \qquad T_{\mu\nu} = \int_0^H {}^\mu U^{(1)\nu}U^{(2)*}\,\mathrm{d}x_3, \tag{4.76} $$

$$ \mu, \nu = 1, 2, \ldots $$

Further let us assume that only the material parameters μ and ϱ are changing at the boundary $x_1 = 0$, and let us define the auxiliary variables η_1 and η_2 by

$$ \eta_1 = 1 - \mathrm{e}^{-2\varepsilon H/\mu_1}, \qquad \eta_2 = 1 - \mathrm{e}^{-2\varepsilon H/\mu_2}. \tag{4.77} $$

To eliminate the sought for reflection and transmission coefficients of the slow mode these are expressed for the present by the coefficients of the fast modes. To this end it is convenient to multiply the equations (4.74) and (4.75) by ${}^0U^{(2)*}$ and ${}^\mu U^{(2)*}$, respectively, and to integrate from 0 to H. By using the orthonormality relations (3.23) and (3.26) as well as the abbreviations (4.76) and (4.77) the four equations

$$ {}^0C^{(2)2} \frac{\mu_2\eta_2}{2\varepsilon} b_0 = (1 + a_0) P + \sum_{\mu=1}^{\infty} a_\mu Q_\mu, \tag{4.78} $$

$$ 2H^\nu C^{(2)2} b_\nu = (1 + a_0) R_\nu + \sum_{\mu=1}^{\infty} a_\mu T_{\mu\nu}, \tag{4.79} $$

$$ \frac{\mu_2^{2\,0}C^{(2)2\,0}k^{(2)}\,\eta_2}{2\mu_1\varepsilon} b_0 = {}^0k^{(1)}(1 - a_0) P - \sum_{\mu=1}^{\infty} a_\mu {}^\mu k^{(1)} Q_\mu, \tag{4.80} $$

$$ 2H^\nu C^{(2)2} \frac{\mu_2}{\mu_1} {}^\nu k^{(2)} b_\nu = {}^0k^{(1)}(1 - a_0) R_\nu - \sum_{\mu=1}^{\infty} a_\mu {}^\mu k^{(1)} T_{\mu\nu} \tag{4.81} $$

are obtained. Note that the orthonormality relations contain the material parameter μ explicitely in conformity with the definition of the scalar product [compare with (2.25)], but here the only product of the corresponding eigenfunctions have been used. This is possible because μ is not a function of x_3 for the model under consideration and, consequently, it cancels on both sides of the equations. By combining the equations (4.78) and (4.80) and after carrying out some transformations, the desired relation for a_0 is obtained as

$$a_0 = \frac{\mu_1 {}^0k^{(1)} - \mu_2 {}^0k^{(2)}}{\mu_1 {}^0k^{(1)} + \mu_2 {}^0k^{(2)}} - \frac{1}{P\left(1 + \dfrac{\mu_1 {}^0k^{(1)}}{\mu_2 {}^0k^{(2)}}\right)} \sum_{\mu=1}^{\infty} a_\mu Q_\mu \left(1 + \frac{\mu_1 {}^\mu k^{(1)}}{\mu_2 {}^0k^{(2)}}\right), \tag{4.82}$$

and by substituting this equation into (4.78) an analogous one follows for b_0:

$$b_0 = \frac{2\varepsilon\mu_1}{{}^0C^{(2)2}\mu_2\eta_2(\mu_1 {}^0k^{(1)} + \mu_2 {}^0k^{(2)})} \left[2 {}^0k^{(1)} P + \sum_{\mu=1}^{\infty} ({}^0k^{(1)} - {}^\mu k^{(1)})\, a_\mu Q_\mu\right]. \tag{4.83}$$

Combining (4.79) and (4.81) and considering (4.82), an infinite system of equations is obtained for the determination of the unknown quantities a_μ ($\mu = 1, 2, \ldots$):

$$\sum_{\mu=1}^{\infty} a_\mu \left\{ \left({}^\nu k^{(2)} + \frac{\mu_1}{\mu_2} {}^\mu k^{(1)}\right) T_{\mu\nu} - \frac{R_\nu \left(1 + \dfrac{\mu_1 {}^\mu k^{(1)}}{\mu_2 {}^0k^{(2)}}\right) \left({}^\nu k^{(2)} + \dfrac{\mu_1}{\mu_2} {}^0k^{(1)}\right)}{P\left(1 + \dfrac{\mu_1 {}^0k^{(1)}}{\mu_2 {}^0k^{(2)}}\right)} Q_\mu \right\}$$

$$+ \frac{2\mu_1 {}^0k^{(1)} ({}^\nu k^{(2)} - {}^0k^{(2)})}{\mu_1 {}^0k^{(1)} + \mu_2 {}^0k^{(2)}} R_\nu = 0, \quad \nu = 1, 2, \ldots \tag{4.84}$$

Finally, rearranging (4.79) yields the transmission coefficients b_ν ($\nu = 1, 2, \ldots$),

$$b_\nu = \frac{1}{2H^\nu C^{(2)2}} \left[(1 + a_0) R_\nu + \sum_{\mu=1}^{\infty} a_\mu T_{\mu\nu}\right], \quad \nu = 1, 2, \ldots, \tag{4.85}$$

if the a_ν ($\nu = 0, 1, 2, \ldots$) are known.

Consequently, for the vertical discontinuity within a closed wave-guide, the crucial step is the solution of the infinite system of equations (4.84). According to Mittra and Lee (1971) the occurence of infinite systems of equations is typical in applying the mode-matching procedure. In general, it is not possible to extract an exact solution of this infinite system of equations, and one is forced to resort to approximation techniques, such as truncation or iteration. However, for a certain class of problems, the associated infinite set of equations does lend itself to an exact solution. But this is not the case for our model. We will solve the system of equations approximately both by a perturbation theory and by reduction. First, the mixed scalar products (4.76) have to be evaluated. To this purpose the corresponding eigenfunctions are substituted into the formulas and the integrations are carried out. After some algebra and by using the abbreviation $\Delta\mu = \mu_2 - \mu_1$ we obtain

$$P = {}^0C^{(1)0}C^{(2)} \frac{\mu_1\mu_2}{\varepsilon(\mu_1 + \mu_2)} \left[1 - \exp\left\{-\frac{\varepsilon H(\mu_1 + \mu_2)}{\mu_1\mu_2}\right\}\right], \tag{4.86}$$

$$Q_\nu = {}^\nu C^{(1)0}C^{(2)} \frac{2\mathrm{i}\varepsilon\nu\pi\mu_2\,\Delta\mu}{H\left(\varepsilon^2 + \frac{\nu^2\pi^2\mu_2^2}{H^2}\right)\left(\varepsilon - \frac{\mathrm{i}\nu\pi\mu_1}{H}\right)} [1 - (-1)^\nu\, \mathrm{e}^{-\varepsilon H/\mu_2}], \tag{4.87}$$

$$R_\nu = {}^0C^{(1)\nu}C^{(2)} \frac{2\mathrm{i}\varepsilon\nu\pi\mu_1\,\Delta\mu}{H\left(\varepsilon^2 + \frac{\nu^2\pi^2\mu_1^2}{H^2}\right)\left(\varepsilon + \frac{\mathrm{i}\nu\pi\mu_2}{H}\right)} [1 - (-1)^\nu\, \mathrm{e}^{-\varepsilon H/\mu_1}], \tag{4.88}$$

$$T_{\mu\nu} = \frac{2\,{}^\mu C^{(1)\nu}C^{(2)}}{\left(\varepsilon - \frac{\mathrm{i}\pi\mu\mu_1}{H}\right)\left(\varepsilon + \frac{\mathrm{i}\pi\nu\mu_2}{H}\right)}$$

$$\times \left\{\delta_{\mu\nu} H \left(\varepsilon^2 + \frac{\nu^2\pi^2\mu_1\mu_2}{H^2}\right) - [1 - (-1)^{\mu+\nu}]\,\varepsilon\nu\,\Delta\mu \left(\frac{1}{\mu+\nu} + \frac{1}{\mu-\nu}\right)\right\},$$

$$\mu, \nu = 1, 2, \ldots \tag{4.89}$$

The infinite system of equations may be rewritten in a more compact manner, where the reflection coefficient of the slow mode a_0 is not "separated":

$$\sum_{\nu=0}^{\infty} A_{\mu\nu} a_\nu = D_\mu, \qquad \mu = 0, 1, 2, \ldots, \tag{4.90}$$

$$A_{\mu\nu} = \left({}^\mu k^{(2)} + \frac{\mu_1}{\mu_2}\,{}^\nu k^{(1)}\right) T_{\nu\mu}, \quad D_\mu = \left(\frac{\mu_1}{\mu_2}\,{}^0k^{(1)} - {}^\mu k^{(2)}\right) R_\mu. \tag{4.91}$$

By using the identities

$$T_{0\nu} \equiv R_\nu, \qquad T_{\nu 0} \equiv Q_\nu, \qquad T_{00} \equiv R_0 \equiv Q_0 \equiv P \tag{4.92}$$

the consistency of (4.90) with (4.84) can be easily demonstrated. It has to be noted that the quantities $T_{0\nu}$, $T_{\nu 0}$, T_{00}, Q_0, and R_0 cannot be obtained by equating the corresponding indices in the formulas (4.87)—(4.89) to zero. The infinite system of equations (4.90) is required for the more detailed analysis of the perturbation theory.

4.3.4. *Solution of the system of equations by perturbation theory*

4.3.4.1. Determination of the reflection coefficients a_ν

To apply a perturbation scheme we assume that the material parameter μ is a weakly varying quantity, i.e.

$$\mu_2 = (1 + \varkappa)\,\mu_1 \quad \text{with} \quad \varkappa = \frac{\Delta\mu}{\mu_1} \ll 1. \tag{4.93}$$

For the present, variations of density ϱ are not taken into account. In the following, all quantities of interest are expanded in terms of the small parameter $\varkappa$. By using the equations (4.88), (4.89), and (4.91) it can be easily seen that the expansion of $A_{\mu\nu}$ has a "$\varkappa$-free" term of 0th order for $\mu = \nu$ only, but there is no such a term in the expansion of D_μ, i.e.

$$A_{\mu\nu} = \delta_{\mu\nu} A^0_{\mu\nu} + A^{\mathrm{I}}_{\mu\nu}\varkappa + A^{\mathrm{II}}_{\mu\nu}\varkappa^2 + O(\varkappa^3), \tag{4.94}$$

$$D_\mu = D^{\mathrm{I}}_\mu \varkappa + D^{\mathrm{II}}_\mu \varkappa^2 + O(\varkappa^3), \qquad \mu, \nu = 0, 1, 2, \ldots \tag{4.95}$$

Further, assuming that for a_ν the expansion

$$a_\nu = a^0_\nu + a^{\mathrm{I}}_\nu \varkappa + a^{\mathrm{II}}_\nu \varkappa^2 + O(\varkappa^3) \tag{4.96}$$

is valid, the system of equations (4.90) may be written as

$$\sum_{\nu=0}^{\infty} [A^0_{\mu\mu} a^0_\mu + (A^0_{\mu\mu} a^{\mathrm{I}}_\mu + A^{\mathrm{I}}_{\mu\nu} a^0_\nu)\,\varkappa + (A^0_{\mu\mu} a^{\mathrm{II}}_\mu + A^{\mathrm{I}}_{\mu\nu} a^{\mathrm{I}}_\nu + A^{\mathrm{II}}_{\mu\nu} a^0_\nu)\,\varkappa^2]$$
$$+ 0(\varkappa^3) = D^{\mathrm{I}}_\mu \varkappa + D^{\mathrm{II}}_\mu \varkappa^2 + O(\varkappa^3), \qquad \mu = 0, 1, 2, \ldots \tag{4.97}$$

A comparison of coefficients of the powers of $\varkappa$ yields to begin with, for the 0th order

$$A^0_{\mu\mu} a^0_\mu = 0, \qquad \mu = 0, 1, 2, \ldots \tag{4.98}$$

Assuming,

$$A^0_{\mu\mu} \neq 0, \qquad \mu = 0, 1, 2, \ldots, \tag{4.99}$$

it follows the vanishing of the "$\varkappa$-free" terms in the expansion of a_ν, i.e.

$$a^0_\nu = 0, \qquad \nu = 0, 1, 2, \ldots \tag{4.100}$$

For the approximate coefficients of 1st and 2nd order, we obtain the expressions

$$a^{\mathrm{I}}_\nu = \frac{D^{\mathrm{I}}_\nu}{A^0_{\nu\nu}}, \tag{4.101}$$

$$a^{\mathrm{II}}_\nu = \frac{1}{A^0_{\nu\nu}} \left(D^{\mathrm{II}}_\nu - \sum_{\mu=0}^{\infty} \frac{A^{\mathrm{I}}_{\nu\mu} D^{\mathrm{I}}_\mu}{A^0_{\nu\nu}} \right), \qquad \nu = 0, 1, 2, \ldots \tag{4.102}$$

Using the expansions

$$T_{\mu\mu} = T^0_{\mu\mu} + T^{\mathrm{I}}_{\mu\mu}\varkappa + O(\varkappa^2), \qquad \mu = 1, 2, \ldots, \tag{4.103}$$

$$T_{0\mu} = \delta_{\mu 0}T^0_{0\mu} + T^{\mathrm{I}}_{0\mu}\varkappa + O(\varkappa^2), \qquad \mu = 0, 1, 2, \ldots, \tag{4.104}$$

$${}^{\mu}k^{(2)} = {}^{\mu}k^{(1)} + {}^{\mu}K^{\mathrm{I}}\varkappa + O(\varkappa^2), \qquad \mu = 0, 1, 2, \ldots, \tag{4.105}$$

and considering (4.91) and (4.92) the quantities D^{I}_{ν} and $A^0_{\nu\nu}$ are obtained as

$$D^{\mathrm{I}}_{\nu} = ({}^0k^{(1)} - {}^{\nu}k^{(1)})\, T^{\mathrm{I}}_{0\nu} - \delta_{\nu 0}({}^0k^{(1)} + {}^{\nu}K^{\mathrm{I}})\, T^0_{0\nu}, \tag{4.106}$$

$$A^0_{\nu\nu} = 2{}^{\nu}k^{(1)}T^0_{\nu\nu}, \qquad \nu = 0, 1, 2, \ldots, \tag{4.107}$$

and it follows according to (4.101) that

$$a^{\mathrm{I}}_{\nu} = \frac{({}^0k^{(1)} - {}^{\nu}k^{(1)})\, T^{\mathrm{I}}_{0\nu} - \delta_{\nu 0}({}^0k^{(1)} + {}^{\nu}K^{\mathrm{I}})\, T^0_{00}}{2{}^{\nu}k^{(1)}T^0_{\nu\nu}}, \qquad \nu = 0, 1, 2, \ldots \tag{4.108}$$

The explicit expressions for T^0_{00}, $T^{\mathrm{I}}_{0\nu}$, $T^0_{\nu\nu}$, and ${}^{\nu}K^{\mathrm{I}}$ are obtained from (3.11), (3.32), (4.86), (4.88), and (4.89) as follows:

$$T^0_{00} = \frac{\mu_1}{2\varepsilon}\left(1 - \mathrm{e}^{-\frac{2\varepsilon H}{\mu_1}}\right) {}^0C^{(1)2}, \tag{4.109}$$

$$T^{\mathrm{I}}_{0\nu} = \frac{2{}^0C^{(1)}\mathrm{i}\varepsilon\nu\pi\mu_1^2}{H\sqrt{2H}\left(\varepsilon^2 + \dfrac{\nu^2\pi^2\mu_1^2}{H^2}\right)\left(\varepsilon + \dfrac{\mathrm{i}\nu\pi\mu_1}{H}\right)} \left[1 - (-1)^{\nu}\,\mathrm{e}^{-\frac{\varepsilon}{\mu_1}H}\right],$$

$$\nu = 1, 2, \ldots, \tag{4.110}$$

$$T^0_{\nu\nu} = 1, \qquad \nu = 1, 2, \ldots, \tag{4.111}$$

$${}^0K^{\mathrm{I}} = -\frac{k^2_{\beta_1} + 2\dfrac{\varepsilon^2}{\mu_1^2}}{2{}^0k^{(1)}}, \qquad {}^{\nu}K^{\mathrm{I}} = -\frac{k^2_{\beta_1}}{2{}^{\nu}k^{(1)}}, \qquad \nu = 1, 2, \ldots, \tag{4.112}$$

so that the reflection coefficients of first approximation are given by

$$a^{\mathrm{I}}_0 = -\frac{k^2_{\beta_1}}{4{}^0k^{(1)^2}} \tag{4.113}$$

and

$$a^{\mathrm{I}}_{\nu} = \frac{({}^0k^{(1)} - {}^{\nu}k^{(1)})\, \mathrm{i}\varepsilon\nu\pi\mu_1^2\left[1 - (-1)^{\nu}\,\mathrm{e}^{-\frac{\varepsilon}{\mu_1}H}\right]{}^0C^{(1)}}{H\sqrt{2H}\left(\varepsilon^2 + \dfrac{\nu^2\pi^2\mu_1^2}{H^2}\right)\left(\varepsilon + \dfrac{\mathrm{i}\nu\pi\mu_1}{H}\right){}^{\nu}k^{(1)}},$$

$$\nu = 1, 2, \ldots \tag{4.114}$$

The expression (4.113) follows immediately from the first term of the right side of (4.82). It has been added for the sake of completeness only. We realize that a_ν^{I} has a singularity (point of resonance) for the limit period T_ν of the ν-th fast mode [compare with (3.12)!], because the vanishing ${}^{\nu}k^{(1)}$ is in the denominator of (4.114). The occurrence of resonances is not unusual in connection with closed wave-guides. For example, TORVIK (1967) found a sharp resonance of a non-propagating mode for the reflection of waves in half-infinite plates.

4.3.4.2. Discussion of the "resonant case" $A_{\nu\nu}^0 = 0$

It follows from equation (4.107) that the condition $A_{\nu\nu}^0 = 0$ is identical to the vanishing of ${}^{\nu}k^{(1)}$. What happens at this point? By looking at the expansion (4.105) of ${}^{\nu}k^{(2)}$, we realize that the expansion coefficients ${}^{\nu}K^{\mathrm{I}}$ become infinite at the critical point $T = T_\nu$, and therefore the expansion (4.105) as well as the expression (4.108) lose their meaning. The mathematical reason for this behaviour is the occurrence of a vertical tangent at the curve

$$ {}^{\nu}k^{(2)} = \sqrt{\frac{\omega^2}{\beta_1^2(1+\varkappa)} - \frac{\nu^2\pi^2}{H^2}} \tag{4.115} $$

for $\varkappa = 0$, which is equivalent with the impossibility of developing the functions $A_{\mu\nu}$ and D_μ in terms of the perturbation parameter $\varkappa$. Hence, this results in the failure of the perturbation method to solve the infinite system of equations at the resonant frequency. Exactly at this point, the respective fast mode changes from the unattenuated into the attenuated propagation. Of course, the reflection coefficients a_ν $(\nu = 1, 2, \ldots)$ are finite for any material contrast and for all periods, but at the critical points it is precisely these which cannot be determined on the basis of this method. It still has to be clarified, how the perturbation method works in the vicinity of the critical points. There, in principle, no anomalous behaviour should occur. The series (4.105) converges for all $T \neq T_\nu$, however, its speed of convergence is crucially dependent on the value of the ratio $\tilde{K}_\nu$,

$$ \tilde{K}_\nu = k_{\beta_1} \cdot \varkappa / {}^{\nu}k^{(1)}, \qquad \nu = 1, 2, \ldots \tag{4.116} $$

If $\tilde{K}^\nu \ll 1$, the series converges quickly and a termination at the first term is generally permissable, i.e. the approximation $a_\nu \approx a_\nu^{\mathrm{I}}\varkappa$ holds. If $\tilde{K}_\nu \gg 1$, we have a completely different situation. Then many terms of higher order are necessary in the series (4.96) to obtain an acceptable approximation for a_ν.

4.3.5. *Determination of the reflection and transmission coefficients for the slow mode with the LQBR-method (Least Square Boundary Residual Method)*

4.3.5.1. Wave-guide type I

While the fundamentals of the LQBR-method will be discussed thoroughly in connection with the open wave-guide, here we treat only the application of this procedure in a modified form for the closed wave-guide. Just the same way as the LQBR-method is able to yield certain information about surface waves for an open wave-guide by neglecting the body waves, here it is possible to obtain information about the slow modes by neglecting the fast ones. Of course, it would also be possible to restrict the infinite number of fast modes and subsequently to apply the LQBR-method and finally to examine the quality of the obtained slow mode coefficients dependent on the number of fast modes taken into account. However, to preserve the analogy with the open wave-guide we neglect here the fast modes completely. This means that in the equations (4.72) and (4.73), the sums are terminated at the term $\nu = 0$ and the missing terms are comprehended in the entities $\tilde{U}$ and $\tilde{S}_{12}$, respectively. With the notation (4.3), we obtain the following system of equations:

$$ {}^0U^{(1)} + a_0 {}^0U^{(1)} = b_0 {}^0U^{(2)} + \tilde{U}, \tag{4.117} $$

$$ {}^0S_{12}^{(1)} - a_0 {}^0S_{12}^{(1)} = b_0 {}^0S_{12}^{(2)} + \tilde{S}_{12}. \tag{4.118} $$

Using the entities $\tilde{U}$ and $\tilde{S}_{12}$ a functional F_2 is constructed [see section 4.4.9.2. and especially (4.290)], whose steadystate condition yields the approximate reflection and transmission coefficients a_0 and b_0 of the slow mode. The procedure is analogous to the open wave-guide, only the integration interval $(0, \infty)$ is replaced by $(0, H)$. Using the notations k_1 and k_2 instead of ${}^0k^{(1)}$ and ${}^0k^{(2)}$ for the sake of brevity, we obtain after some algebra

$$ a_0^{\mathrm{LQ}} = \frac{(\mu_1 k_1 - \mu_2 k_2)(\mu_1 k_1 + \mu_2 k_2)}{\dfrac{16}{\omega^2 P^2} + (\mu_1 k_1 - \mu_2 k_2)^2}, \tag{4.119} $$

$$ b_0^{\mathrm{LQ}} = \frac{4(\mu_1 k_1 + \mu_2 k_2)}{\dfrac{16}{\omega P} + (\mu_1 k_1 - \mu_2 k_2)^2 \,\omega P}. \tag{4.120} $$

If the integration constants are specified for unit energy transport as we have arranged, and using (4.86), P is found as

$$ P = \frac{4\left[1 - \exp\left\{-\dfrac{\varepsilon H(\mu_1 + \mu_2)}{\mu_1 \mu_2}\right\}\right]}{\omega(\mu_1 + \mu_2)\sqrt{k_1 k_2 \eta_1 \eta_2}}. \tag{4.121} $$

For comparison, the expressions (4.119) and (4.120) are calculated by the computer program CLWG (see section 4.3.6.).

4.3.5.2. Wave-guide type II

For comparison we present here formulas being analogous to (4.119) and (4.120), because an exact solution can be found for the wave-guide type II using the Wiener-Hopf technique (see section 5.3.), as long as the discontinuity is caused by a jump of the surface impedance $\varepsilon_1 \to \varepsilon_2$ only. By using the abbreviations

$$P_1 = \frac{C_1^2}{2\gamma_1}(1 - e^{-2\gamma_1 H}), \quad P_2 = \frac{C_2^2}{2\gamma_2}(1 - e^{-2\gamma_2 H}), \tag{4.122}$$

$$P_{12} = \frac{C_1 C_2}{\gamma_1 + \gamma_2}[1 - e^{-(\gamma_1+\gamma_2)H}], \tag{4.123}$$

where γ_1 and γ_2 are solutions of the equations

$$\tanh(\gamma_1 H) = \frac{\varepsilon_1}{\mu\gamma_1} \quad \text{and} \quad \tanh(\gamma_2 H) = \frac{\varepsilon_2}{\mu\gamma_2}, \tag{4.124}$$

respectively, [compare with (3.16) and (5.114)], these formulas are given by

$$a_0^{LQ} = \frac{k_1^2 - k_2^2}{4k_1k_2 \frac{P_1P_2}{P_{12}^2} + (k_1 - k_2)^2}, \tag{4.125}$$

$$b_0^{LQ} = \frac{2(k_1 + k_2)\, P_1}{4k_1k_2 \frac{P_1P_2}{P_{12}} + (k_1 - k_2)^2 P_{12}}. \tag{4.126}$$

In section 4.4.9.4., we will compare the approximation (4.125) with the exact reflection coefficient after Wiener-Hopf.

4.3.6. *Solution of the system of equations using a computer program and comparison of the various solutions*

To obtain a solution independent of the requirements of the perturbation method, it is necessary to solve the infinite system of equations (4.90) by reduction, that means the number of the used modes is limited to $\nu_{\max}$, i.e. $\nu = 0, 1, 2, \ldots, \nu_{\max}$. We hope that the solution obtained in this way converges rapidly enough to the exact one. For this purpose a computer program CLWG was written in FORTRAN for the computer BESM-6, whose basic subroutine inverts the matrix $A_{\mu\nu}$, which is complex in general, in double precision. The program CLWG calculates the real and imaginary parts of all a_ν and b_ν ($\nu = 1$, $2, \ldots, \nu_{\max}$) by reduction, the real and imaginary parts of a_ν^I ($\nu = 1, 2, \ldots, \nu_{\max}$), as well as the LQBR-approximation of a_0 and b_0 and their deviations. Further, a quantity PEUF (Percentage of Energy Unaccounted For) is calculated to control the energy balance. Let us proceed from the assumption that the energy is distributed between the individual partial waves in accordance with the normalization of section 3.2.4. as follows:

energy of the incident slow mode = 1,

energy of the reflected slow mode = $|a_0|^2$,

energy of the transmitted slow mode = $|b_0|^2$,

energy of the reflected (unattenuated) fast modes

$= \frac{\omega}{2} \mu_1 {}^{\nu}k^{(1)} |a_\nu|^2, \ \nu = 1, \ldots, \nu_{\max},$

energy of the transmitted (unattenuated) fast modes

$$= \frac{\omega}{2} \mu_2 {}^{\nu}k^{(2)} |b_\nu|^2, \ \nu = 1, \ldots, \nu_{\max}. \tag{4.127}$$

Then PEUF, defined by

$$\text{PEUF} = 100(1 - |a_0|^2 - |b_0|^2 - \frac{\omega}{2} \mu_1 \sum_{\nu=1}^{M_u} {}^{\nu}k^{(1)} |a_\nu|^2 - \frac{\omega}{2} \mu_2 \sum_{\nu=1}^{N_u} {}^{\nu}k^{(2)} |a_\nu|^2), \tag{4.128}$$

is a quantity, which has to vanish in the ideal case. Here M_u and N_u are the numbers of unattenuated fast modes in the media 1 and 2, respectively,

$$M_u = \left[\frac{Hk_{\beta_1}}{\pi}\right], \quad N_u = \left[\frac{Hk_{\beta_2}}{\pi}\right] \quad ([\cdots]: \text{read greatest whole number}). \tag{4.129}$$

Obviously, the attenuated fast modes as well as the "critical modes" with ${}^{\nu}k^{(1)} = 0$ or ${}^{\nu}k^{(2)} = 0$ do not contribute to the energy balance. The vanishing of PEUF is equivalent to the fulfilment of the energy conservation law. The obtained coefficients a_ν and b_ν are not exact, because $\nu_{\max}$ is always finite

Table 3. Model parameters for the closed wave-guide

Model	μ_1 10^9 N/m²	μ_2 10^9 N/m²	ϱ_1 g/cm³	ϱ_2 g/cm³	ε 10^6 N/m³	H km	$\nu_{\max}$
16/17-1	16	17	4	4	10	1	10
16/17-100	16	17	4	4	10	100	10
75/100-100	75	100	3	3	10	100	20

(it is assumed $\nu_{max} > M_u$ and $\nu_{max} > N_u$), and consequently PEUF $\neq 0$. Although we will treat PEUF as a certain quality parameter for our calculations, we must state in the final analysis that its vanishing is necessary, of course, but not sufficient for exact values of a_ν and b_ν. Table 3 summarizes the parameters of the calculated models.

Both the models 16/17-1 and 16/17-100 are characterized by a slight μ-contrast but they have different thicknesses. Let a characteristic parameter $\tilde{p}$ be the ratio of the wave length of the slow mode to the thickness of the waveguide,

$$\tilde{p} = \lambda_0/H. \tag{4.130}$$

For the considered period range, this parameter varies from 0.2 to 2.7 for model 16/17-1 and from 0.09 to 0.1 for model 16/17-100. In Fig. 20, we see the absolute values of the reflection coefficients a_ν with $\nu = 0, 1, 2, 4$ calculated

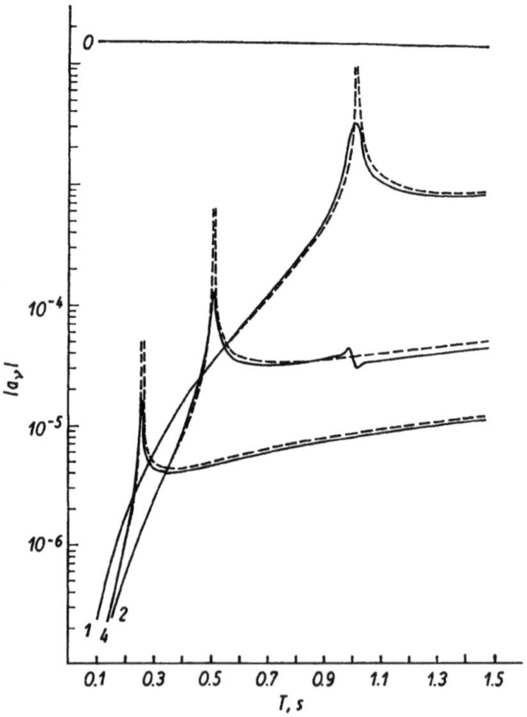

Fig. 20. The reflection coefficients $|a_\nu|$ ($\nu = 0, 1, 2, 4$) versus period for model 16/17-1

full lines: solutions by reduction (CLWG),

dashed lines: approximations according to (4.114)

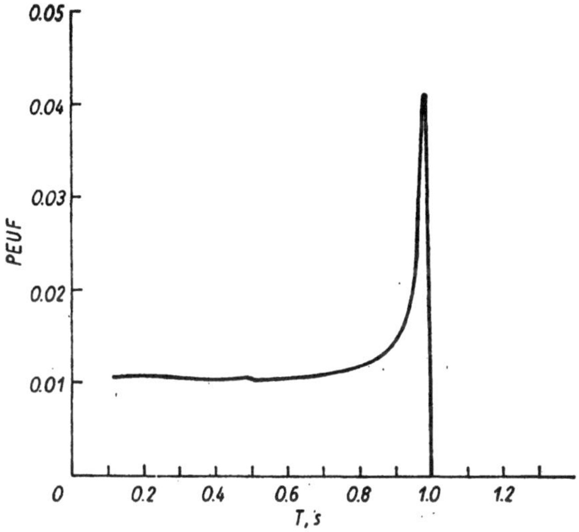

Fig. 21. PEUF versus period for model 16/17-1

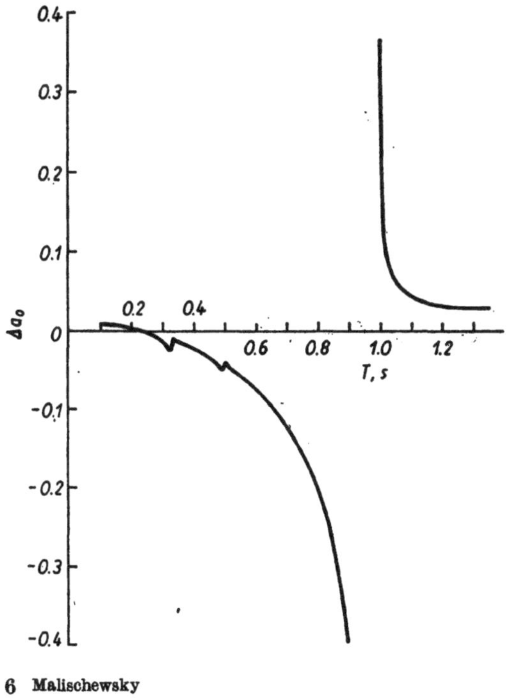

Fig. 22. Percentage deviation Δa_0 of $|a_0^{LQ}|$ from $|a_0|$ versus period for model 16/17-1

by CLWG and represented logarithmically in dependence of the period T for model 16/17-1 (full lines).

The dashed lines represent the approximations $|a_\nu^{\mathrm{I}}| \cdot \varkappa$ for the fast modes according to (4.114). For the considered period range, the reflection coefficient $|a_0|$ shows a very slight dependence on period only, and the given curve practically coincides with the LQBR-approximation according to (4.119) [see Fig. 22]. The strong dependence on period for the reflection coefficients of fast modes and especially the maxima of those at the resonance points $T_1 = 1$ s, $T_2 = 0.5$ s, and $T_4 = 0.25$ s are worth noting. As it was to be expected, the curves based on the reduction procedure remain finite at the critical points. However, the dashed curves diverge at these points, but otherwise they coincide very well with the CLWG-curves. We have seen that the perturbation method fails in these points due to the impossibility of developing certain functions. It is possible to carry out a more heuristic consideration by separating the corresponding expressions in developable and non-developable parts and, subsequently, by developing the first ones only according to the perturbation schema. The approximations obtained in this way coincide comparatively well with the maxima from CLWG. This confirms our former conception. The quantities PEUF and the percentage deviation Δa_0 of $|a_0^{\mathrm{LQ}}|$ from $|a_0|$ are plotted versus period in Fig. 21 and Fig. 22, respectively.

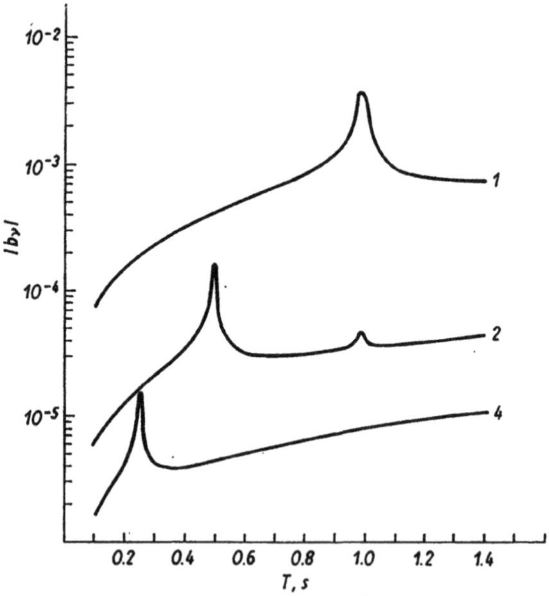

Fig. 23. The transmission coefficients $|b_\nu|$ ($\nu = 1, 2, 4$) versus period for model 16/17-1

Note that $M_u = 0$ for $T > 1$ s according to (4.129), i.e. all reflected fast modes are attenuated in medium 1 in this period range.

In Fig. 23 the absolute values of the transmission coefficients $|b_\nu|$ with $\nu = 1, 2, 4$ calculated by CLWG are plotted logarithmically versus period for model 16/17-1. The transmission coefficient $|b_0|$ is as good as equal to 1 in the whole period range under consideration. The deviation from the similarly calculated LQBR-approximation is not greater than 0.005%.

Note again the maxima of the fast modes at critical periods T_ν. The maxima of the reflection as well as transmission coefficients may be explained by the fact that in the vicinity of the critical points the energy transport of the corresponding modes is very small, but on the other hand the energy balance should be ensured.

The following figures refer to model 16/17-100, at which now $T_1 = 100$ s, $T_2 = 50$ s, and $T_4 = 25$ s hold. Fig. 24 corresponds to Fig. 20.

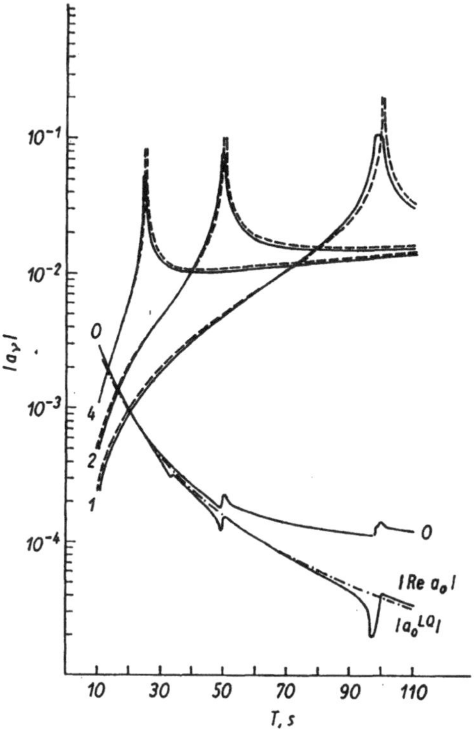

Fig. 24. The reflection coefficients $|a_\nu|$ ($\nu = 0, 1, 2, 4$) versus period for model 16/17-100

thin: $|\mathrm{Re}\, a_0|$, dash-dotted: $|a_0^{LQ}|$

The agreement between the approximations $|a_\nu^{\mathrm{I}}| \cdot \varkappa$ and $|a_\nu|$ is again very good with the exception of the critical points. However, the LQBR-approximation of $|a_0|$ has drastically deteriorated, especially for greater periods. Let us give some comments on this behaviour. Fig. 24 additionally contains the absolute value of the real part of a_0 according to CLWG. Obviously, this agrees comparatively well with $|a_0^{\mathrm{LQ}}|$ in the whole period range under consideration with the exception of the resonance points which "penetrate" from the fast modes into the slow one. On the other hand, a_0^{LQ} agrees very well with $a_0^{\mathrm{I}} \cdot \varkappa$. As a result, we can proceed on the assumption that the real part of a_0 is approached on the whole by the first approximation (4.113) and the imaginary part by the second [see (4.102)] and higher approximations, respectively. For certain $\bar{p}$-ratios it is quite possible that in the series (4.96) the term $a_0^{\mathrm{II}} \cdot \varkappa^2$ exceeds $a_0^{\mathrm{I}} \cdot \varkappa$. Of course, this is independent of the fact that the series converges ultimately. It follows that in these cases the LQBR-approximation will be very bad. This corresponds with the statement of MALISCHEWSKY

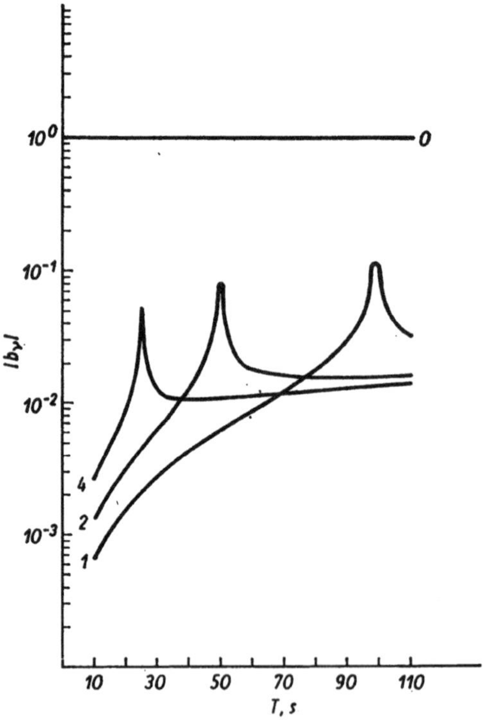

Fig. 25. The transmission coefficients $|b_\nu|$ ($\nu = 0, 1, 2, 4$) versus period for model 16/17-100

(1976) that Alsop's method, based on the LQBR-method, is unsuitable for determining imaginary parts of reflection and transmission coefficients of surface waves.

PEUF is not greater than 0.1% in almost the whole range. In Fig. 25 the absolute values of the transmission coefficients $|b_\nu|$ ($\nu = 0, 1, 2, 4$) are plotted versus period for model 16/17-100. The percentage deviation of the almost constant coefficient b_0 from b_0^{LQ} is less than 0.01% in the whole period range.

Finally we consider the model 75/100-100 with a clearly higher material contrast, and that is why we have doubled ν_{max}, too. The characteristic results of the reflection coefficients are represented in Fig. 26.

Firstly it is noticed that the agreement between $|a_\nu|$ and $|a_\nu^I| \cdot \varkappa$ ($\nu = 1, 2, 4$) is poorer in comparison with the models 16/17-1 and 16/17-100. This is not surprising, because $\varkappa \approx 0.06$ for the above-mentioned models, but $\varkappa = 1/3$ for model 75/100-100. Further we notice that, taken absolutely, a_0 is smaller

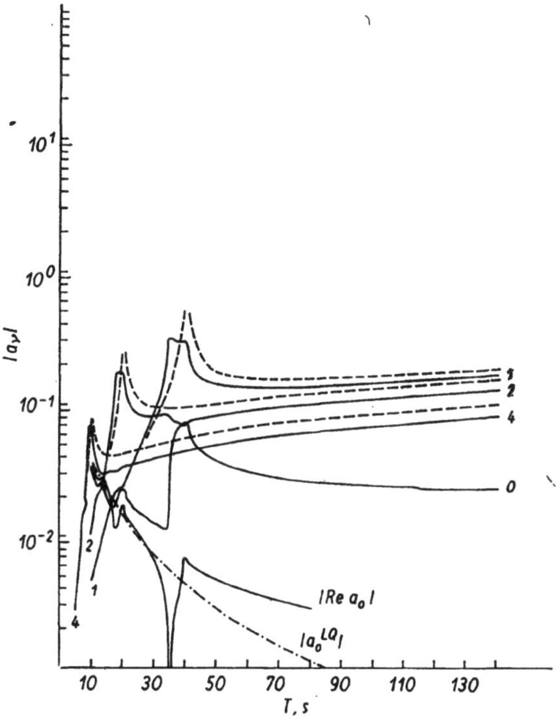

Fig. 26. The reflection coefficients $|a_\nu|$ ($\nu = 0, 1, 2, 4$) versus period for model 75/100-100

than a_ν ($\nu = 1, 2, 4$) in nearly the whole range. This indicates that with increasing material contrast at the discontinuity higher modes are increasingly excited, which is quite plausible. For $T > 15$ s, there is no more a correspondence between $|a_0|$ and $|a_0^{LQ}|$, but $|a_0^{LQ}|$ still approximates the curve $|\mathrm{Re}\, a_0|$ quite well for $T \leqq 30$ s. However, $|b_0|$ is approximated by $|b_0^{LQ}|$ with a deviation of less than 1% in the whole range under consideration. In this range, PEUF is less than 0.1%.

On the whole, our theoretical conceptions have been confirmed by these calculations. Critical quantities are the material contrast and the parameter $\tilde{p}$. For the reflection coefficient of the slow mode, the quality of the LQBR-approximation depends crucially on these quantities. The transmission coefficient of the slow mode is a much less sensitive quantity in comparison, which in all cases is approximated very well by the LQBR-method.

4.3.7. *Comparison of the exact solution after Wiener-Hopf with the LQBR-approximation concerning the reflection coefficient of the slow mode for a wave-guide type II with an impedance jump*

According to (5.123) the absolute value of the reflection coefficient for the slow mode is given by the simple formula

$$|R_0| = \frac{|k_1 - k_2|}{k_1 + k_2}, \tag{4.131}$$

where the wave numbers k_1 and k_2 are understood again as refering to the slow mode in medium 1 and 2, respectively. The following Table 4 shows $|R_0|$, a_0^{LQ}, and the percentage error of a_0^{LQ}, $\delta\%$, dependent on $\varkappa_1$ for variable H, ε and T, where in accordance with (4.254) $\varkappa_1$ is defined by

$$\varepsilon_2 = \varkappa_1 \varepsilon_1 \overset{!}{=} \varkappa_1 \varepsilon. \tag{4.132}$$

The parameters μ and ϱ have been fixed as $\mu = 18 \cdot 10^9$ N/m^2 and $\varrho = 2$ g/cm^3. For the variables, the same dimensions are valid as in Table 3. It has still to be noted that formula (4.131) only represents the absolute value of the reflection coefficient, if merely the slow mode can propagate in medium 1, i.e. if according to (3.17)

$${}^1\gamma_1^2 > k_\beta^2, \qquad \frac{\pi}{2H} \leqq {}^1\gamma_1 \leqq \frac{\pi}{H} \tag{4.133}$$

holds. This condition is fulfilled for our parameter combinations.

Table 4. Comparison of Wiener-Hopf and LQBR-solutions for various combinations

$\varkappa$	$\lvert R_0\rvert$	a_0^{LQ}	$\lvert\delta\%\rvert$	$\lvert R_0\rvert$	a_0^{LQ}	$\lvert\delta\%\rvert$	$\lvert R_0\rvert$	a_0^{LQ}	$\lvert\delta\%\rvert$
0.1	0.1750	0.1746	0.2	0.0318	0.0310	2.5	0.5268	0.5259	0.2
0.3	0.1233	0.1232	0.09	0.0246	0.0243	1.1	0.3008	0.3005	0.1
0.5	0.0815	0.0814	0.04	0.0178	0.0177	0.5	0.1826	0.1825	0.04
0.7	0.0446	0.0446	0.01	0.0095	0.0095	0.1	0.0991	0.0991	0.02
0.9	0.0148	0.0148	0.001	0.0049	0.0049	0.03	0.0265	0.0265	0.001
1.1	0.0147	−0.0147	0.001	0.0052	−0.0052	0.03	0.0127	−0.0127	0.0003
1.3	0.0366	−0.0366	0.01	0.0107	−0.0106	0.1	0.0606	−0.0606	0.01
1.5	0.0651	−0.0651	0.03	0.0223	−0.0222	0.4	0.0935	−0.0935	0.02
1.7	0.0859	−0.0858	0.05	0.0284	−0.0282	0.7	0.1243	−0.1242	0.04
2.0	0.1128	−0.1127	0.08	0.0479	−0.0471	1.7	0.1711	−0.1710	0.08
5.0	0.3001	−0.2980	0.7	0.2192	−0.1822	16.9	0.3924	−0.3898	0.7
10.0	0.4546	−0.4451	2.1	0.4514	−0.2747	39.2	0.5458	−0.5357	1.9
	$H = 1,\ \varepsilon = 1,\ T = 10$			$H = 10,\ \varepsilon = 1,\ T = 10$			$H = 1,\ \varepsilon = 1,\ T = 100$		
	$\lvert F_e\rvert = 0.07$			$\lvert F_e\rvert = 2.07$			$\lvert F_e\rvert = 0.07$		

$\varkappa$	$\lvert R_0\rvert$	a_0^{LQ}	$\lvert\delta\%\rvert$	$\lvert R_0\rvert$	a_0^{LQ}	$\lvert\delta\%\rvert$
0.1	0.4534	0.4439	2.1	0.6517	0.5561	14.7
0.3	0.2859	0.2827	1.1	0.4440	0.3969	10.6
0.5	0.1783	0.1774	0.5	0.2904	0.2736	5.8
0.7	0.0973	0.0971	0.2	0.1611	0.1578	2.1
0.9	0.0293	0.0293	0.02	0.0489	0.0488	0.2
1.1	0.0278	−0.0278	0.02	0.0466	−0.0465	0.2
1.3	0.0762	−0.0761	0.2	0.1271	−0.1252	1.5
1.5	0.1202	−0.1197	0.4	0.1959	−0.1889	3.5
1.7	0.1603	−0.1591	0.7	0.2552	−0.2401	5.9
2.0	0.2131	−0.2101	1.4	0.3292	−0.2977	9.6
5.0	0.5356	−0.4662	13.0	0.6636	−0.4613	30.5
10.0	0.7352	−0.5372	26.9	0.8164	−0.4903	39.9
	$H = 1,\ \varepsilon = 10,\ T = 10$			$H = 1,\ \varepsilon = 50,\ T = 10$		
	$\lvert F_e\rvert = 0.21$			$\lvert F_e\rvert = 0.43$		

From Table 4 the following trends emerge:

1. The period T influences the magnitude of the reflection coefficient more than the error $\delta\% = \left|\dfrac{R_0 - a_0^{LQ}}{R_0}\right| \cdot 100.$
2. For fixed T and ε $\delta\%$ increases with increasing H.
3. For fixed T and H $\delta\%$ increases with increasing ε.

The behaviour according to 1. can be understood if we consider that T does not influence the amplitude-depth functions as opposed to surface waves in layered media. Further it has to be noted that a suitable quality test for the LQBR-method is missing. A quantity F_e, defined by

$$F_e = \frac{\gamma_2(1 - e^{-\gamma_1 H}) - \gamma_1(1 - e^{-\gamma_2 H})}{\gamma_1\gamma_2}, \tag{4.134}$$

could yield a certain indication in this direction. It is the difference area formed by the two amplitude-depth functions and is included in Table 4. We have determined a certain proportionality between the error magnitude and F_e.

4.4. The open wave-guide with a vertical discontinuity and an impedance jump

4.4.1. *General remarks*

As already mentioned, the procedure is in general such that we try to apply the methods developed for the closed wave-guide with disturbances to the open wave-guide by modifying them in some way. For the various methods this is successful in different ways and to varying degrees. The correspondence between closed and open wave-guides has been summarized in Table 2. In response to this correspondence Brooke and Kharadly (1976) suggested the following procedure: Ideally conductive boundaries are introduced in an open electromagnetic wave-guide, subsequently the methods of the closed wave-guide are applied and, finally, the solutions obtained are interpreted as approximate solutions for the open wave-guide. According to these authors it is possible to obtain quite good approximations for the open configuration in this manner, even for reasonable distances of the boundaries from the original guiding structure. This statement was confirmed by the paper of Stegeman et al. (1981) as already mentioned. On the other hand, it is not clear whether we always obtain satisfying results in this manner. After all, the radiation conditions are changed considerably by it [see also the remark at Rozzi (1978)]. For this reason, in the following we shall try to treat the problem of discontinuities in an open wave-guide directly. The difficulties have caused many authors to concentrate upon the comparatively simple dielectric slab wave-guide, which is, without doubt, an idealization of the problems appearing in practice. An exceptionally simple dieletric slab wave-guide consists of a dielectric layer adjoining to the vacuum. It is well-known, that such a structure can guide surface waves corresponding to Love waves in the simple model "layer over half-space" in the elastic case. In this connection there is a correspondence between the elastic half-space and the vacuum. If the electric properties are changing jump — like in the direction of propagation of the waves, we have to deal with a fundamental type of discontinuities in an open electromagnetic wave-guide. The seismological analogue would be a discontinuity with a finite extension in depth, and that is that case in reality we are actually interested in. We will go into details in section 4.5.3.3. Nevertheless we will also consider such vertical discontinuities which extend up to infinity. There are two reasons for doing so. On the one hand, we have to proceed from the assumption that finitely and infinitely deep discontinuities cannot practically be distinguished for small periods. On the other hand, from a historical point of view an attempt was first of all made to treat the exceptionally simple model of two unlayered half-spaces welded together, where the discontinuity is automatically extended up to infinity. Of course, the great simplicity of that model was achieved at the expense of having to consider the effect of this discontinuity on the more complicated Rayleigh waves, because there are no Love waves in such a structure. By introducing the conception of the so-called impedance surface we will avoid this difficulty, as already mentioned. We obtain the impedance surface by reducing the thickness of the slab wave-guide up to zero and by simultaneously fulfilling certain boundary conditions (see section 2.1.2.). Assuming an impedance jump at a fixed place and considering the behaviour of SH-like surface waves ("Love-kind" surface waves) in this structure, as will be done in the following, we have a situation which in the main corresponds to the study of an infinitely extended vertical discontinuity. It is true that discontinuities with finite extension in depth cannot be treated in this way, but a crucial problem concerning the interaction of surface waves with discontinuities can be investigated in detail when one has at hand an exact solution according to Wiener-Hopf.

The step discontinuity (see Fig. 1a) is a further fundamental disturbance, which may occur in a slab wave-guide. It is treated in modified form by using the same approximate methods as for vertical or sloping discontinuities. Although the step discontinuity, perhaps in combination with vertical discontinuities, is also very interesting from the geophysical point of view, we will focus our attention on impedance jumps and vertical discontinuities with finite or infinite extension in depth alone. The step discontinuities have merely been taken into account in the references.

4.4.2. *Application of the mode-matching method*

In the theory of wave-guides, one of the most important but also most difficult questions deals with completeness of the eigenfunction system of the wave-guide. Let this question be solved positively, and let us represent any field in the wave-guide in the form of a spectral expansion in terms of this set of orthogonal eigenfunctions. The eigenfunctions of the open wave-guide having a

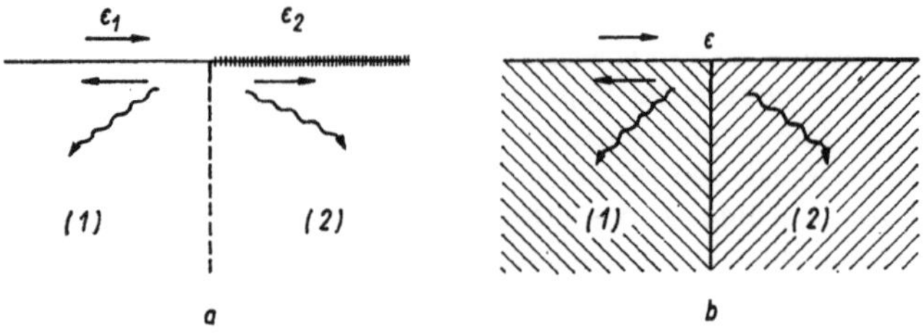

Fig. 27. The disturbances under consideration in the open wave-guide
a) impedance jump, b) vertical discontinuity

discrete and continuous eigenspectrum have been calculated in section 3.3.4. Assuming a geometry shown in Fig. 27, let us proceed analogously to the closed wave-guide (section 4.3.2.).

Instead of sums [compare with (4.67)—(4.70)] we now have integrals over the continuous eigenfunctions, and the resulting displacements and stresses in the regions (1) and (2) are obtained as

$$u_{\text{res}}^{(1)} = {}^0U^{(1)}(x_3)\, e^{i\,{}^0k^{(1)}x_1} + a_0\,{}^0U^{(1)}(x_3)\, e^{-i\,{}^0k^{(1)}x_1} + \int_0^\infty a(\gamma)\, U^{(1)}(x_3, \gamma)\, e^{-i\sqrt{k_{\beta_1}^2 - \gamma^2}\,x_1}\, d\gamma, \tag{4.135}$$

$$u_{\text{res}}^{(2)} = b_0\,{}^0U^{(2)}(x_3)\, e^{i\,{}^0k^{(2)}x_1} + \int_0^\infty b(\gamma)\, U^{(2)}(x_3, \gamma)\, e^{i\sqrt{k_{\beta_2}^2 - \gamma^2}\,x_1}\, d\gamma, \tag{4.136}$$

$$\sigma_{12\text{res}}^{(1)} = \mu_1 i\,{}^0k^{(1)}\,{}^0U^{(1)}(x_3)\, e^{i\,{}^0k^{(1)}x_1} - a_0 i\mu_1\,{}^0k^{(1)}\,{}^0U^{(1)}(x_3)\, e^{-i\,{}^0k^{(1)}x_1} - i\mu_1 \int_0^\infty a(\gamma) \sqrt{k_{\beta_1}^2 - \gamma^2}\, U^{(1)}(x_3, \gamma)\, e^{-i\sqrt{k_{\beta_1}^2 - \gamma^2}\,x_1}\, d\gamma, \tag{4.137}$$

$$\sigma_{12\text{res}}^{(2)} = \mu_2 i\,{}^0k^{(2)} b_0\,{}^0U^{(2)}(x_3)\, e^{i\,{}^0k^{(2)}x_1} + i\mu_2 \int_0^\infty b(\gamma) \sqrt{k_{\beta_2}^2 - \gamma^2}\, U^{(2)}(x_3, \gamma)\, e^{i\sqrt{k_{\beta_2}^2 - \gamma^2}\,x_1}\, d\gamma. \tag{4.138}$$

The quantities a_0 and b_0 are generalized reflection and transmission coefficients of surface waves, respectively (in our simple wave-guide there is only one discrete surface wave mode at all!). On the other hand, the "continuous expansion coefficients" $a(\gamma)$ and $b(\gamma)$ characterize that part of the incident surface waves converted into radiation modes or body waves and scattered into the regions (1) and (2), respectively.

The fulfilment of the continuity conditions (4.71) leads to the system of equations

$$
{}^0U^{(1)}(x_3) + a_0{}^0U^{(1)}(x_3) + \int_0^\infty a(\gamma)\, U^{(1)}(x_3, \gamma)\, \mathrm{d}\gamma
$$

$$
= b_0{}^0U^{(2)}(x_3) + \int_0^\infty b(\gamma)\, U^{(2)}(x_3, \gamma)\, \mathrm{d}\gamma, \tag{4.139}
$$

$$
{}^0k^{(1)0}U^{(1)}(x_3) - a_0{}^0k^{(1)0}U^{(1)}(x_3) - \int_0^\infty a(\gamma) \sqrt{k_{\beta_1}^2 - \gamma^2}\, U^{(1)}(x_3, \gamma)\, \mathrm{d}\gamma
$$

$$
= \frac{\mu_2}{\mu_1} b_0{}^0k^{(2)0}U^{(2)}(x_3) + \frac{\mu_2}{\mu_1} \int_0^\infty b(\gamma) \sqrt{k_{\beta_2}^2 - \gamma^2}\, U^{(2)}(x_3, \gamma)\, \mathrm{d}\gamma. \tag{4.140}
$$

Using the bra-ket formalism, it is possible to write the system of equations in a considerably more compact form. Denoting the resulting field vector with $\Psi_{1\mathrm{res}}$ and $\Psi_{2\mathrm{res}}$ in the regions (1) and (2), respectively, and analogously the ψ-vectors according to (4.4), the equations (4.135)—(4.138) are obtained as

$$
\Psi_{1\mathrm{res}} = |\psi_1^{(+)}\rangle\, \mathrm{e}^{\mathrm{i}^0k^{(1)}x_1} + a_0\, |\psi_1^{(-)}\rangle\, \mathrm{e}^{-\mathrm{i}^0k^{(1)}x_1}
$$

$$
+ \int_0^\infty a(\gamma)\, |\psi_1^{(-)}(\gamma)\rangle\, \mathrm{e}^{-\mathrm{i}\sqrt{k_{\beta_1}^2 - \gamma^2}x_1}\, \mathrm{d}\gamma, \tag{4.141}
$$

$$
\Psi_{2\mathrm{res}} = b_0\, |\psi_2^{(+)}\rangle\, \mathrm{e}^{\mathrm{i}^0k^{(2)}x_1} + \int_0^\infty b(\gamma)\, |\psi_2^{(+)}(\gamma)\rangle\, \mathrm{e}^{\mathrm{i}\sqrt{k_{\beta_2}^2 - \gamma^2}x_1}\, \mathrm{d}\gamma, \tag{4.142}
$$

and the continuity conditions are simply

$$
\Psi_{1\mathrm{res}} = \Psi_{2\mathrm{res}} \quad \text{for} \quad x_1 = 0, \tag{4.143}
$$

and it follows

$$
|\psi_1^{(+)}\rangle + a_0\, |\psi_1^{(-)}\rangle + \int_0^\infty a(\gamma)\, |\psi_1^{(-)}(\gamma)\rangle\, \mathrm{d}\gamma = b_0\, |\psi_2^{(+)}\rangle + \int_0^\infty b(\gamma)\, |\psi_2^{(+)}(\gamma)\rangle\, \mathrm{d}\gamma. \tag{4.144}
$$

4.4.3. *Setting up of a singular integral equation for the determination of the reflection and transmission coefficients*

Taking the scalar product of (4.144) with $\langle\psi_2^{(-)}|$, $\langle\psi_2^{(-)}(\sigma)|$ for $\sigma < k_{\beta_2}$ and with $\langle\psi_2^{(+)}(\sigma)|$ for $\sigma > k_{\beta_2}$, respectively, where Λ was identified with k_{β_2} [compare with (4.21)], we obtain successively the equations

$$\langle\psi_2^{(-)} \mid \psi_1^{(+)}\rangle + a_0\langle\psi_2^{(-)} \mid \psi_1^{(-)}\rangle + \int_0^\infty a(\gamma)\,\langle\psi_2^{(-)} \mid \psi_1^{(-)}(\gamma)\rangle\,\mathrm{d}\gamma = 0, \tag{4.145}$$

$$\langle\psi_2^{(-)}(\sigma) \mid \psi_1^{(+)}\rangle + a_0\langle\psi_2^{(-)}(\sigma) \mid \psi_1^{(-)}\rangle + \int_0^\infty a(\gamma)\,\langle\psi_2^{(-)}(\sigma) \mid \psi_1^{(-)}(\gamma)\rangle\,\mathrm{d}\gamma = 0$$
$$\text{for} \quad \sigma < k_{\beta_2}, \tag{4.146}$$

$$\langle\psi_2^{(+)}(\sigma) \mid \psi_1^{(+)}\rangle + a_0\langle\psi_2^{(+)}(\sigma) \mid \psi_1^{(-)}\rangle + \int_0^\infty a(\gamma)\,\langle\psi_2^{(+)}(\sigma) \mid \psi_1^{(-)}(\gamma)\rangle\,\mathrm{d}\gamma = 0$$
$$\text{for} \quad \sigma > k_{\beta_2}; \tag{4.147}$$

from which an integral equation for $a(\gamma)$ alone may be obtained. For the simple case under consideration we will not continue using the bra-ket formalism, but the integral equation will be derived directly from the equations (4.139) and (4.140). Because there is one surface wave mode only, we will mostly omit the index "0". Further, in many cases the x_3-dependence of the eigenfunctions will not be written explicitly. Analogously to (4.76), let us introduce the scalar products

$$P = \int_0^\infty U^{(1)}U^{(2)*}\,\mathrm{d}x_3, \qquad Q(\gamma) = \int_0^\infty U^{(2)*}U^{(1)}(\gamma)\,\mathrm{d}x_3,$$

$$R(\gamma) = \int_0^\infty U^{(1)}U^{(2)*}(\gamma)\,\mathrm{d}x_3, \tag{4.148}$$

$$T(\gamma, \sigma) = \int_0^\infty U^{(1)}(\gamma)\,U^{(2)*}(\sigma)\,\mathrm{d}x_3.$$

In equations (4.139) and (4.140) the quantities a, b, $a(\gamma)$ and $b(\gamma)$ are unknowns. We propose to determine $a(\gamma)$ and then to derive the other quantities from it. The elimination of $a(\gamma)$ is carried out using the same scheme as described in the important paper of Rulf and Kedem (1976). For this purpose the equations (4.139) and (4.140) are multiplied successively with $U^{(2)*}$ and $U^{(2)*}(\sigma)$ and then they are integrated over x_3 from 0 to ∞. Using the orthonormality relations (3.68), (3.69), (3.75), and the abbreviations (4.148) the following system of integral equations is obtained for the unknowns a, b, $a(\gamma)$, and $b(\gamma)$:

$$ {}^{0}C^{(2)2}\,\frac{\mu_2}{2\varepsilon}\,b = (1 +, a)\,P + \int_0^\infty a(\gamma)\,Q(\gamma)\,d\gamma\,, \tag{4.149}$$

$$ 2\pi C^{(2)2} b(\gamma) = (1 + a)\,R(\gamma) + \int_0^\infty a(\sigma)\,T(\sigma,\gamma)\,\mathrm{d}\sigma\,, \tag{4.150}$$

$$ \frac{\mu_2^{2\,0}C^{(2)2}k^{(2)}}{2\mu_1\varepsilon}\,b = k^{(1)}(1 - a)\,P - \int_0^\infty a(\gamma)\sqrt{k_{\beta_1}^2 - \gamma^2}\,Q(\gamma)\,\mathrm{d}\gamma\,, \tag{4.151}$$

$$ 2\pi C^{(2)2}\,\frac{\mu_2}{\mu_1}\sqrt{k_{\beta_2}^2 - \gamma^2}\,b(\gamma) = k^{(1)}(1 - a)\,R(\gamma) - \int_0^\infty a(\sigma)\sqrt{k_{\beta_1}^2 - \sigma^2}\,T(\sigma,\gamma)\,\mathrm{d}\gamma\,. \tag{4.152}$$

From the equations (4.149) and (4.151), a can be eliminated in the manner

$$ a = \frac{\mu_1 k^{(1)} - \mu_2 k^{(2)}}{\mu_1 k^{(1)} + \mu_2 k^{(2)}} - \frac{1}{P\left(1 + \dfrac{\mu_1 k^{(1)}}{\mu_2 k^{(2)}}\right)} \times \int_0^\infty a(\gamma)\,Q(\gamma)\left(1 + \frac{\mu_1}{\mu_2 k^{(2)}}\sqrt{k_{\beta_1}^2 - \gamma^2}\right)\mathrm{d}\gamma\,, \tag{4.153}$$

and using (4.150), (4.152), and (4.153) we obtain an integral equation for $a(\gamma)$ alone:

$$ \frac{2\mu_1 k_1\left(\sqrt{k_{\beta_2}^2 - \gamma^2} - k_2\right)}{\mu_1 k_1 + \mu_2 k_2}\,R(\gamma) + \int_0^\infty a(\sigma)\left\{\left(\sqrt{k_{\beta_2}^2 - \gamma^2} + \frac{\mu_1}{\mu_2}\sqrt{k_{\beta_1}^2 - \sigma^2}\right)T(\sigma,\gamma) - \frac{R(\gamma)\left(1 + \dfrac{\mu_1}{\mu_2 k_2}\sqrt{k_{\beta_1}^2 - \sigma^2}\right)\left(\sqrt{k_{\beta_2}^2 - \gamma^2} + \dfrac{\mu_1 k_1}{\mu_2}\right)}{P\left(1 + \dfrac{\mu_1 k_1}{\mu_2 k_2}\right)}\,Q(\sigma)\right\}\mathrm{d}\sigma = 0. \tag{4.154}$$

If this integral equation can be solved, the other unknown quantities a, b, and $b(\gamma)$ can be obtained by integrations of known functions. First the scalar products (4.148) have to be evaluated. We proceed from a vertical discontinuity with jump-like change of the material parameters μ and ϱ. As for the closed wave-guide, the corresponding eigenfunctions are substituted into the formulas and the integrations are carried out. Using the abbreviation $\Delta\mu = \mu_2 - \mu_1$ we obtain after some algebra

$$P = {}^0C^{(1)0}C^{(2)} \frac{\mu_1\mu_2}{\varepsilon(\mu_1 + \mu_2)}, \tag{4.155}$$

$$Q(\gamma) = {}^0C^{(2)}C^{(1)} \frac{2i\varepsilon\gamma\mu_2 \, \Delta\mu}{(\varepsilon^2 + \gamma^2\mu_2^2)(\varepsilon - i\gamma\mu_1)}, \tag{4.156}$$

$$R(\gamma) = {}^0C^{(1)}C^{(2)} \frac{2i\varepsilon\gamma\mu_1 \, \Delta\mu}{(\varepsilon^2 + \gamma^2\mu_1^2)(\varepsilon + i\gamma\mu_2)}. \tag{4.157}$$

The evaluation of $T(\gamma, \sigma)$ is more difficult and we want to represent it here in detail. Denoting for a short time

$$\frac{\varepsilon + i\gamma\mu}{\varepsilon - i\gamma\mu} = e^{i\varphi(\gamma)} \tag{4.158}$$

the continuous eigenfunctions are

$$U(x_3, \gamma) = C\{e^{i\gamma x_3} - e^{i[\varphi(\gamma) - \gamma x_3]}\} \tag{4.159}$$

and the "mixed product" of the eigenfunctions is

$$\begin{aligned} &U^{(1)}(x_3, \gamma)\, U^{(2)*}(x_3, \sigma) \\ &= C^{(1)}C^{(2)}\{e^{i(\gamma-\sigma)x_3} - e^{-i\varphi_2(\sigma)}e^{i(\gamma+\sigma)x_3} - e^{i\varphi_1(\gamma)}e^{-i(\gamma+\sigma)x_3} \\ &\quad + e^{i[\varphi_1(\gamma)-\varphi_2(\sigma)]}e^{-i(\gamma-\sigma)x_3}\}. \end{aligned} \tag{4.160}$$

Integrating this relationship over x_3 from 0 to ∞ and applying formula (3.72), $T(\gamma, \sigma)$ is obtained as

$$\begin{aligned} \frac{1}{C^{(1)}C^{(2)}} \cdot T(\gamma, \sigma) = \pi\{1 + e^{i[\varphi_1(\gamma)-\varphi_2(\sigma)]}\} \\ \times \delta(\gamma - \sigma) - \pi\{e^{i\varphi_1(\gamma)} + e^{-i\varphi_2(\sigma)}\}\, \delta(\gamma + \sigma) \\ + i\{1 - e^{i[\varphi_1(\gamma)-\varphi_2(\sigma)]}\}\, P\frac{1}{\gamma - \sigma} \\ - i\{e^{-i\varphi_2(\gamma)} - e^{i\varphi_1(\sigma)}\}\, P\frac{1}{\gamma + \sigma}. \end{aligned} \tag{4.161}$$

Assuming $\gamma + \sigma \neq 0$ the term with $\delta(\gamma + \sigma)$ does not have to be considered. The expressions in the other braces are given by

$$\begin{aligned} 1 + e^{i[\varphi_1(\gamma) - \varphi_2(\sigma)]} &= \frac{2(\varepsilon^2 + \gamma\sigma\mu_1\mu_2)}{(\varepsilon - i\gamma\mu_1)(\varepsilon + i\sigma\mu_2)}, \\ 1 - e^{i[\varphi_1(\gamma) - \varphi_2(\sigma)]} &= \frac{2i\varepsilon(\sigma\mu_2 - \gamma\mu_1)}{(\varepsilon - i\gamma\mu_1)(\varepsilon + i\sigma\mu_2)}, \\ e^{-i\varphi_2(\sigma)} - e^{i\varphi_1(\gamma)} &= -\frac{2i\varepsilon(\gamma\mu_1 + \sigma\mu_2)}{(\varepsilon - i\gamma\mu_1)(\varepsilon + i\sigma\mu_2)}, \end{aligned} \tag{4.162}$$

and finally it follows that

$$\begin{aligned} T(\gamma, \sigma) = {} & \frac{2C^{(1)}C^{(2)}}{(\varepsilon - i\gamma\mu_1)(\varepsilon + i\sigma\mu_2)} \\ & \times \left[\pi(\varepsilon^2 + \gamma\sigma\mu_1\mu_2)\,\delta(\gamma - \sigma)\right. \\ & \left. - \varepsilon\,\Delta\mu\sigma\left(\mathrm{P}\frac{1}{\gamma - \sigma} + \mathrm{P}\frac{1}{\gamma + \sigma}\right)\right] \quad \text{for} \quad \gamma + \sigma \neq 0. \end{aligned} \tag{4.163}$$

Because the formulas are needed later on, let us evaluate here the scalar products for the generalized case "vertical discontinuity + impedance jump" too, i.e. for a combination of the discontinuities according to Fig. 27a and b. Denoting the impedance parameter left and right of the discontinuity with ε_1 and ε_2, respectively, and the impedance jump with $\Delta\varepsilon = \varepsilon_2 - \varepsilon_1$, we obtain analogous to (4.155)—(4.157), and (4.163)

$$P = {}^0C^{(1)}\,{}^0C^{(2)}\,\frac{\mu_1\mu_2}{\varepsilon_1\mu_2 + \varepsilon_2\mu_1}, \tag{4.164}$$

$$Q(\gamma) = {}^0C^{(2)}C^{(1)}\,\frac{2i\gamma\mu_2(\varepsilon_1\,\Delta\mu - \mu_1\,\Delta\varepsilon)}{(\varepsilon_2^2 + \gamma^2\mu_2^2)(\varepsilon_1 - i\gamma\mu_1)}, \tag{4.165}$$

$$R(\gamma) = {}^0C^{(1)}C^{(2)}\,\frac{2i\gamma\mu_1(\varepsilon_1\,\Delta\mu - \mu_1\,\Delta\varepsilon)}{(\varepsilon_1^2 + \gamma^2\mu_1^2)(\varepsilon_2 + i\gamma\mu_2)}, \tag{4.166}$$

$$\begin{aligned} T(\gamma, \sigma) = {} & \frac{2C^{(1)}C^{(2)}}{(\varepsilon_1 - i\gamma\mu_1)(\varepsilon_2 + i\sigma\mu_2)} \\ & \times \left[\pi(\varepsilon_1\varepsilon_2 + \gamma\sigma\mu_1\mu_2)\,\delta(\gamma - \sigma) - \varepsilon_1\,\Delta\mu\sigma\right. \\ & \times \left(\mathrm{P}\frac{1}{\gamma - \sigma} + \mathrm{P}\frac{1}{\gamma + \sigma}\right) - \mu_1\,\Delta\varepsilon\gamma \\ & \left. \times \left(\mathrm{P}\frac{1}{\gamma + \sigma} - \mathrm{P}\frac{1}{\gamma - \sigma}\right)\right] \\ & \text{for} \quad \gamma + \sigma \neq 0. \end{aligned} \tag{4.167}$$

Substituting (4.155)—(4.157), and (4.163) into (4.154) the integral equation for $a(\gamma)$ can be written as

$$\frac{2\mathrm{i}\,{}^0C^{(1)}\varepsilon\mu_1\,\Delta\mu\gamma\left(\sqrt{k_{\beta_2}^2-\gamma^2}-k_2\right)}{\left(1+\frac{\mu_2k_2}{\mu_1k_1}\right)(\varepsilon^2+\gamma^2\mu_1^2)}$$

$$+\frac{\pi C^{(1)}\left(\sqrt{k_{\beta_2}^2-\gamma^2}+\frac{\mu_1}{\mu_2}\sqrt{k_{\beta_1}^2-\gamma^2}\right)(\varepsilon^2+\gamma^2\mu_1\mu_2)}{(\varepsilon-\mathrm{i}\gamma\mu_1)}\,a(\gamma)$$

$$-\varepsilon\,\Delta\mu\gamma\int\limits_0^\infty\frac{C^{(1)}}{\varepsilon-\mathrm{i}\sigma\mu_1}\left\{\left(\sqrt{k_{\beta_2}^2-\gamma^2}+\frac{\mu_1}{\mu_2}\sqrt{k_{\beta_1}^2-\sigma^2}\right)\mathrm{P}\,\frac{1}{\sigma+\gamma}\right.$$

$$+\frac{2\varepsilon^2\,\Delta\mu(\mu_1+\mu_2)\left(\sqrt{k_{\beta_2}^2-\gamma^2}+\frac{\mu_1k_1}{\mu_2}\right)}{(\varepsilon^2+\gamma^2\mu_1^2)\left(1+\frac{\mu_1k_1}{\mu_2k_2}\right)}$$

$$\left.\times\frac{\sigma\left(1+\frac{\mu_1}{\mu_2k_2}\sqrt{k_{\beta_1}^2-\sigma^2}\right)}{(\varepsilon^2+\sigma^2\mu_2^2)}\right\}a(\sigma)\,\mathrm{d}\sigma$$

$$-\Delta\mu\gamma\int\limits_0^\infty\frac{C^{(1)}\left(\sqrt{k_{\beta_2}^2-\gamma^2}+\frac{\mu_1}{\mu_2}\sqrt{k_{\beta_1}^2-\sigma^2}\right)}{(\varepsilon-\mathrm{i}\sigma\mu_1)}\,\mathrm{P}\,\frac{1}{\sigma-\gamma}\,a(\sigma)\,\mathrm{d}\sigma=0. \tag{4.168}$$

Before we discuss this equation let us also quote the explicit formulas for the other unknowns. Substituting (4.155) and (4.156) into (4.153) the reflection coefficient of surface waves is found as

$$a=\frac{\mu_1k_1-\mu_2k_2}{\mu_1k_1+\mu_2k_2}-\frac{2\mathrm{i}\varepsilon^2\,\Delta\mu(\mu_1+\mu_2)\,C^{(1)}}{{}^0C^{(1)}\mu_1\left(1+\frac{\mu_1k_1}{\mu_2k_2}\right)}\int\limits_0^\infty\frac{\gamma\left(1+\frac{\mu_1}{\mu_2k_2}\sqrt{k_{\beta_1}^2-\gamma^2}\right)}{(\varepsilon^2+\gamma^2\mu_2^2)(\varepsilon-\mathrm{i}\gamma\mu_1)}\,a(\gamma)\,\mathrm{d}\gamma, \tag{4.169}$$

and using (4.149) the transmission coefficient of surface waves is

$$b = \frac{4\mu_1 k_1 {}^0C^{(1)}}{{}^0C^{(2)}\left(1 + \frac{\mu_2}{\mu_1}\right)(\mu_1 k_1 + \mu_2 k_2)} + \frac{4\mathrm{i}\varepsilon^2 \mu_1 \Delta\mu C^{(1)}}{{}^0C^{(2)}(\mu_1 k_1 + \mu_2 k_2)}$$

$$\times \int_0^\infty \frac{\gamma\left(k_1 + \sqrt{k_{\beta_1}^2 - \gamma^2}\right)}{(\varepsilon^2 + \gamma^2\mu_2^2)\cdot(\varepsilon - \mathrm{i}\gamma\mu_1)}\, a(\gamma)\, \mathrm{d}\gamma. \tag{4.170}$$

By using (4.157) and (4.163) the amplitude coefficient $b(\gamma)$ of the body waves scattered into medium 2 follows from (4.150) as

$$b(\gamma) = \frac{{}^0C^{(1)}\mathrm{i}\varepsilon\gamma\mu_1\Delta\mu(1 + a)}{\pi C^{(2)}(\varepsilon^2 + \gamma^2\mu_1^2)(\varepsilon + \mathrm{i}\gamma\mu_2)} + \frac{C^{(1)}(\varepsilon^2 + \gamma^2\mu_1\mu_2)\, a(\gamma)}{C^{(2)}(\varepsilon - \mathrm{i}\gamma\mu_1)(\varepsilon + \mathrm{i}\gamma\mu_2)}$$

$$- \frac{C^{(1)}\varepsilon\Delta\mu\gamma}{\pi C^{(2)}(\varepsilon + \mathrm{i}\gamma\mu_2)} \int_0^\infty \frac{\left(\mathrm{P}\frac{1}{\sigma - \gamma} + \mathrm{P}\frac{1}{\sigma + \gamma}\right)}{\varepsilon - \mathrm{i}\sigma\mu_1}\, a(\sigma)\, \mathrm{d}\sigma, \tag{4.171}$$

where we have not substituted a according to (4.169) for the sake of clarity. In the following, we will study more thoroughly $a(\gamma)$ and a only.

4.4.4. *Discussion of the singular integral equation for $a(\gamma)$*

The amplitude coefficient $a(\gamma)$ of the backwards (i.e. into region 1) scattered body waves satisfies an integral equation of the type

$$f(\gamma) + g(\gamma)\, a(\gamma) + \int_0^\infty H(\gamma, \sigma)\, a(\sigma)\, \mathrm{d}\sigma + \int_0^\infty \frac{K(\sigma, \gamma)}{\sigma - \gamma}\, a(\sigma)\, \mathrm{d}\sigma = 0. \tag{4.172}$$

That is an inhomogeneous singular integral equation of the second kind with an infinite integration interval, where the integral with the essential singular kernel $K(\sigma, \gamma)/(\sigma - \gamma)$ is to be understood in the Cauchy sense. The solution of this integral equation is determined essentially by the asymptotic behaviour of as well $f(\gamma)$, $g(\gamma)$, $H(\gamma, \sigma)$, $K(\gamma, \sigma)$ as well as the function sought for $a(\gamma)$ itself. The solutions of integral equations are sought for in certain function spaces suited to the problem. Here we are particularly interested in the spaces of continuous functions and quadratically integrable functions. First let us consider the limit $\lim_{\gamma \to 0} a(\gamma)$. It follows from equation (3.64) that the continuous eigenfunctions used by us vanish for $\gamma \to 0$, so that $\lim_{\gamma \to 0} a(\gamma)$ is still undetermined.

Incidently, this is not the case for the continuous eigenfunctions used by RULF and KEDEM (1976), which are defined by

$$U_{\mathrm{RULF}} = \cos \gamma x_3 - \frac{\varepsilon}{\gamma\mu} \sin \gamma x_3 = \frac{\gamma\mu + i\varepsilon}{2\mu\gamma} \cdot U(x_3, \gamma). \tag{4.173}$$

These eigenfunctions have the limit

$$\lim_{\gamma \to 0} U_{\mathrm{RULF}} = 1 - \frac{\varepsilon}{\mu} x_3, \tag{4.174}$$

so that it would be necessary to require $a(0) = 0$ in order to avoid an infinite growth for $x_3 \to \infty$. The appearing of eigenfunctions with a linear x_3-dependence is connected with the existence of so-called Goodier-Bishop waves [see MALISCHEWSKY (1971)]. For the sake of completeness we quote here the continuous eigenfunctions used by SHEVCHENKO (1971a) too:

$$U_{\mathrm{SEV}} = \gamma\mu \cos \gamma x_3 - \varepsilon \sin \gamma x_3 = \frac{\gamma\mu + i\varepsilon}{2} \cdot U(x_3, \gamma). \tag{4.175}$$

From the integral equation (4.168) it follows immediately that $a(0) = 0$. But we should mention the following difficulty. By carrying out the limiting process $\gamma \to 0$ directly at Q, R, and T under the integral sign and subsequently substituting the results into the integral equation (4.154), we obtain the trivial representation $0 = 0$, but no statement about $a(0)$. That is because of the fact that for improper integrals depending on a parameter it is not always permitted to exchange the limiting process with the integral. For consistency we have to assume $a(0) = 0$ in each case.

Further, let us consider $\lim_{\gamma \to \infty} a(\gamma)$. Obviously, it has to be $a(\infty) = 0$. It is only a question of how rapidly $a(\gamma)$ tends to zero. Using the Dirichlet test, we have to assume an asymptotic behaviour of the form $a(\gamma) = O(1/\gamma^{1+s})$ with $s > 0$ for $\gamma \to \infty$ to ensure the boundedness of the third term on the left side of (4.172), if $a(\gamma)$ does not oscillate continuously. The same asymptotic behaviour also follows from the boundedness of the field vector for $x_1, x_3 \to 0$ (*edge condition*!). A more detailed functional analytic treatment of the eigenvalue problem by SHEVCHENKO (1979) reveals the estimation

$$a(\gamma) = O\left(\frac{1}{\gamma^2}\right) \quad \text{for} \quad \gamma \to \infty, \tag{4.176}$$

for the interval $(-\infty, \infty)$. But as already mentioned, the results can be transferred to the semi-open interval $[0, \infty)$.

It is possible to solve integral equations of the type (4.172) by using numerical as well as analytical methods. An extensive theory was developed by MUSKHELISHVILI (1965). However, the regularization of a singular integral equation is a complicated procedure on this basis, and only for a very limited number of integral equations can an explicit solution be found. Among these there is the integral equation for the impedance jump problem, which is obtained by setting $\Delta\mu = 0$ in the scalar products (4.164)—(4.167) and substituting into (4.154). On this basis, a solution was given e.g. by RULF (1977a) using the canonical function of the corresponding homogeneous equation. But the form of this solution is not convenient for further considerations. We will not study this variant, because that problem has an explicit solution on the basis of Wiener-Hopf. For more general discontinuities we have, in any case, to proceed on the assumption that there are no explicit solutions for the corresponding integral equations and numerical methods have to be used. Among numerous publications concerning the numerical solution of singular integral equations, let us quote the following papers of recent years DOW (1977), COHEN (1978), DOW and ELLIOTT (1979), CARON (1981), and IOAKIMIDIS (1981) which are interesting from the author's view point.

In general, the above-mentioned papers refer to the interval $[-1, 1]$ and not to the semi-infinite one $[0, \infty)$ required by us. However, using the variable transformation

$$\sigma = \frac{1+t}{1-t}, \qquad d\sigma = \frac{2}{(1-\gamma)^2}\,\mathrm{d}t \tag{4.177}$$

it is possible to map the latter onto the interval $[-1, 1]$ in a simple manner. Using the abbreviations

$$\begin{aligned} \gamma &= \frac{1+\tau}{1-\tau}, \qquad & f(\gamma) &= \bar{f}(\tau), \qquad & g(\gamma) &= \bar{g}(\tau), \\ a(\gamma) &= \bar{a}(\tau), & H(\gamma, \sigma) &= \bar{H}(\tau, t), & K(\sigma, \gamma) &= \bar{K}(t, \tau) \end{aligned} \tag{4.178}$$

equation (4.172) appears as

$$\bar{f}(\tau) + \bar{g}(\tau)\,\bar{a}(\tau) + \int_{-1}^{1} \bar{H}(\tau, t)\,\frac{2}{(1-t)^2}\,\bar{a}(t)\,\mathrm{d}t + (1-\tau)\int_{-1}^{1} \frac{\bar{K}(t, \tau)}{t-\tau}\,\frac{\bar{a}(t)}{1-t}\,\mathrm{d}t = 0, \tag{4.179}$$

having already a certain similarity with the equations investigated by DOW (1977) and COHEN (1978)

$$\bar{g}_1(\tau)\,\bar{a}(\tau) + \frac{h(\tau)}{\pi}\int_{-1}^{1} \frac{\bar{a}(t)}{t-\tau}\,\mathrm{d}t + \int_{-1}^{1} \bar{K}_1(\tau, t)\,\bar{a}(t)\,\mathrm{d}t = \bar{f}(\tau),$$

(DOW, complete equation), (4.180)

$$\bar{a}(\tau) = \bar{f}_2(\tau) + \lambda\int_{-1}^{1} \frac{\bar{K}_2(\tau, t)}{\tau - t}\,\bar{a}(t)\,\mathrm{d}t \quad \text{(COHEN)}. \tag{4.181}$$

However, by the transformation an additional difficulty has appeared in the form of a singularity at the end point $t = 1$ of the interval [see (4.179)]. This contradicts the required properties of the integral kernels. According to Cohen (1978) it is also possible to apply the interpolation scheme developed by him directly to an infinite interval without transforming onto the interval $[-1, 1]$. But then, due to numerical reason, it is required that $a(\gamma) = O(e^{-\gamma})$ for $\gamma \to \infty$ holds. Such a behaviour cannot be assumed without further consideration, because up to now only the estimation (4.176) has been available. According to Cohen (1978) the regularization of equation (4.181) is carried out by using the identity

$$\bar{a}(\tau) = \bar{f}_2(\tau) + \lambda \int_{-1}^{1} \frac{[\bar{K}_2(\tau, t) - \bar{K}_2(\tau, \tau)]}{\tau - t} \bar{a}(t) \, dt$$

$$+ \lambda \bar{K}_2(\tau, \tau) \int_{-1}^{1} \bar{a}(t) \, \mathrm{P} \frac{1}{\tau - t} \, dt. \tag{4.182}$$

The first integrand is no longer singular at $t = \tau$ since $\bar{K}_2$ is assumed as "well-behaved" and thus it has a Taylor expansion in t around τ. This means that the first integral can be well approximated by a quadrature sum. The second integral is approximated using Legendre's polynomials, where $\bar{a}(t)$ is evaluated at the same quadrature points at those used to approximate the first integral. Finally, a matrix equation is found for (4.182), which can be solved effectively by matrix inversion.

To avoid the difficulty with the singularity at $t = 1$, the following procedure would be imaginable: Split the integrals as $\int_0^\infty = \int_0^A + \int_A^\infty$, treat the integral $\int_0^A$ according to Cohen, where the mapping of $[0, A]$ onto $[-1, 1]$ succeeds without additional singularities, and integrate the remaining integral using an asymptotic estimation for $a(\gamma)$, e.g. $a(\gamma) = \bar{\varkappa}/\gamma^2$. Let $\bar{\varkappa}$ be a parameter which has to be obtained somehow from the integral equation. The described procedure should lead to numerically stable solutions of the singular integral equation (4.168), however, in practice it has not yet been applied up to now.

It should be mentioned that the method of Caron (1981), based on a projection method in the Hilbert space, was also used for solving wave-guide problems [see Caron et al. (1980)].

A first step in obtaining a general solution of (4.168) may be a particular solution for small jumps of material parameters at the discontinuity, which is researched in the next section. This solution requires attention in itself as well as for the control of any numerical solutions.

4.4.5. *Determination of an approximate solution using a perturbation method*

Singular integral equations being characteristic for discontinuities in wave-guides have been solved for different configurations already by RULF (1976, 1977c) using an asymptotic approach. Let us treat our problem concerning a discontinuity of material parameters μ and ϱ in a similar manner by expressing all quantities with the medium index 2 in terms of quantities with the index 1. Assuming

$$\frac{\Delta\mu}{\mu_1} \ll 1 \quad \text{and} \quad \frac{\Delta\varrho}{\varrho_1} \ll 1 \tag{4.183}$$

series expansions are carried out in $\Delta\mu$ and $\Delta\varrho$. Using the abbreviations

$$\begin{aligned}
\xi_1 &= \left(\frac{\partial\sqrt{k_\beta^2 - \gamma^2}}{\partial\mu}\right)_{\mu_1, \varrho_1}, & \xi_2 &= \left(\frac{\partial\sqrt{k_\beta^2 - \gamma^2}}{\partial\varrho}\right)_{\mu_1, \varrho_1}, \\
\xi_3 &= \left(\frac{\partial k}{\partial\mu}\right)_{\mu_1, \varrho_1}, & \xi_4 &= \left(\frac{\partial k}{\partial\varrho}\right)_{\mu_1, \varrho_1}, \\
\delta_1 &= \left(\frac{\partial U}{\partial\mu}\right)_{\mu_1, \varrho_1}, & \delta_2 &= \left(\frac{\partial U}{\partial\varrho}\right)_{\mu_1, \varrho_1}, \\
\eta(\gamma) &= \left(\frac{\partial U(\gamma)}{\partial\mu}\right)_{\mu_1, \varrho_1}, & \Delta &= \mathrm{Max}\,(|\Delta\mu|, |\Delta\varrho|)
\end{aligned} \tag{4.184}$$

we obtain

$$k_2 = k_1 + \xi_3\Delta\mu + \xi_4\Delta\varrho + O(\Delta^2), \tag{4.185}$$

$$\sqrt{k_{\beta_2}^2 - \gamma^2} = \sqrt{k_{\beta_1}^2 - \gamma^2} + \xi_1\Delta\mu + \xi_2\Delta\varrho + O(\Delta^2), \tag{4.186}$$

$$U^{(2)} = U^{(1)} + \delta_1\Delta\mu + \delta_2\Delta\varrho + O(\Delta^2), \tag{4.187}$$

$$U^{(2)}(\gamma) = U^{(1)}(\gamma) + \eta\Delta\mu + O([\Delta\mu]^2). \tag{4.188}$$

Note that the density dependence of the discrete eigenfunctions follows from the k-dependence of the normalization constant according to (3.76), whereas the continuous eigenfunctions have no density dependence at all.

Denoting

$$\begin{aligned}
P_0 &= \int_0^\infty U^{(1)}U^{(1)*}\,\mathrm{d}x_3, & P_1 &= \int_0^\infty \delta_1 U^{(1)}\,\mathrm{d}x_3, \\
P_2 &= \int_0^\infty \delta_2 U^{(1)}\,\mathrm{d}x_3, & R_1(\gamma) &= \int_0^\infty \eta(\gamma)^*\,U^{(1)}\,\mathrm{d}x_3, \\
Q_1(\sigma) &= \int_0^\infty \delta_1 U^{(1)}(\sigma)\,\mathrm{d}x_3, & Q_2(\sigma) &= \int_0^\infty \delta_2 U^{(1)}(\sigma)\,\mathrm{d}x_3, \\
T_1(\gamma, \sigma) &= \int_0^\infty \eta(\sigma)^*\,U^{(1)}(\gamma)\,\mathrm{d}x_3
\end{aligned} \tag{4.189}$$

and substituting the expansions (4.187) and (4.188) into the scalar products (4.148), the relations

$$P = P_0 + P_1 \Delta\mu + P_2 \Delta\varrho + O(\Delta^2), \tag{4.190}$$

$$R(\gamma) = R_1(\gamma)\, \Delta\mu + O([\Delta\mu]^2), \tag{4.191}$$

$$Q(\sigma) = Q_1(\sigma)\, \Delta\mu + Q_2(\sigma)\, \Delta\varrho + O(\Delta^2), \tag{4.192}$$

$$T(\gamma, \sigma) = \delta(\gamma - \sigma) + T_1(\gamma, \sigma)\, \Delta\mu + O([\Delta\mu]^2) \tag{4.193}$$

are obtained. In deriving (4.191), the orthogonality between discrete and continuous eigenfunctions has been taken into account. In equation (4.193) the normalization constant of the continuous eigenfunctions was fixed according to (3.77). Proceeding from the integral equation (4.154) in a sligthly modified notation

$$\begin{aligned} &2\mu_1\mu_2k_1P\left(\sqrt{k_{\beta_2}^2 - \gamma^2} - k_2\right) R(\gamma) \\ &+ \int_0^\infty a(\sigma) \left\{P(\mu_1k_1 + \mu_2k_2) \left(\mu_1 \sqrt{k_{\beta_1}^2 - \sigma^2} + \mu_2 \sqrt{k_{\beta_2}^2 - \gamma^2}\right)\right. \\ &\times T(\sigma, \gamma) - R(\gamma) \left(\mu_2k_2 + \mu_1 \sqrt{k_{\beta_1}^2 - \sigma^2}\right) \\ &\times \left.\left(\mu_1k_1 + \mu_2 \sqrt{k_{\beta_2}^2 - \gamma^2}\right) Q(\sigma)\right\} \mathrm{d}\sigma = 0 \end{aligned} \tag{4.194}$$

and substituting the expansions (4.190)—(4.193) into this equation, we obtain after some transformations

$$\begin{aligned} &2\mu_1^2k_1P_0 \left(\sqrt{k_{\beta_1}^2 - \gamma^2} - k_1\right) R_1(\gamma)\, \Delta\mu + O(\Delta^2) \\ &+ \int_0^\infty a(\sigma) \left\{\left[2\mu_1^2k_1\left(\sqrt{k_{\beta_1}^2 - \sigma^2} + \sqrt{k_{\beta_1}^2 - \gamma^2}\right) P_0 + (\ldots)\, \Delta\mu\right.\right. \\ &+ \left.(\ldots)\, \Delta\varrho\right] \delta(\gamma - \sigma) + 2\mu_1^2k_1P_0\left(\sqrt{k_{\beta_1}^2 - \gamma^2} + \sqrt{k_{\beta_1}^2 - \sigma^2}\right) \\ &\times \left.T_1(\sigma, \gamma)\, \Delta\mu + O(\Delta^2)\right\} \mathrm{d}\sigma = 0, \end{aligned} \tag{4.195}$$

and finally,

$$\begin{aligned} &2\mu_1^2k_1P_0\left(\sqrt{k_{\beta_1}^2 - \gamma^2} - k_1\right) R_1(\gamma)\, \Delta\mu + O(\Delta^2) \\ &+ \left[4\mu_1^2k_1P_0 \sqrt{k_{\beta_1}^2 - \gamma^2} + (\ldots)\, \Delta\mu + (\ldots)\, \Delta\varrho\right] a(\gamma) \\ &+ \int_0^\infty a(\sigma) \left\{2\mu_1^2k_1P_0\left(\sqrt{k_{\beta_1}^2 - \gamma^2} + \sqrt{k_{\beta_1}^2 - \sigma^2}\right)\right. \\ &\times \left.T_1(\sigma, \gamma)\, \Delta\mu + O(\Delta^2)\right\} \mathrm{d}\sigma = 0. \end{aligned} \tag{4.196}$$

Taking into account the expansion of $a(\gamma)$,

$$a(\gamma) = a^{\mathrm{I}}(\gamma)\, \Delta\mu + \bar{a}^{\mathrm{I}}(\gamma)\, \Delta\varrho + O(\Delta^2), \tag{4.197}$$

which is suggested by the structure of the integral equation (4.194), we conclude initially

$$\bar{a}^{\mathrm{I}}(\gamma) \equiv 0, \tag{4.198}$$

and, finally, the first approximation of the reflection coefficient for radiation modes or body waves is obtained as

$$a^{\mathrm{I}}(\gamma) = \frac{\left(k_1 - \sqrt{k_{\beta_1}^2 - \gamma^2}\right)}{2\sqrt{k_{\beta_1}^2 - \gamma^2}}\, R_1(\gamma). \tag{4.199}$$

For the determination of $R_1(\gamma)$ we need $\eta(\gamma)^*$, which is given by

$$\eta(\gamma)^* = \frac{2\mathrm{i}\gamma\varepsilon C^{(1)}\, \mathrm{e}^{\mathrm{i}\gamma x_3}}{(\varepsilon + \mathrm{i}\gamma\mu_1)^2}, \tag{4.200}$$

and using the normalization (3.76) and (3.77) it follows from (4.189)

$$R_1(\gamma) = \frac{4\mathrm{i}\gamma\varepsilon}{(\varepsilon^2 + \gamma^2\mu_1^2)(\varepsilon + \mathrm{i}\gamma\mu_1)} \sqrt{\frac{\varepsilon}{2\pi\omega k_1}}, \tag{4.201}$$

so that the first approximation (4.199) can be written as

$$\begin{aligned} a^{\mathrm{I}}(\gamma) &= \frac{2\mathrm{i}\gamma\varepsilon\left(k_1 - \sqrt{k_{\beta_1}^2 - \gamma^2}\right)}{(\varepsilon^2 + \gamma^2\mu_1^2)(\varepsilon + \mathrm{i}\gamma\mu_1)\sqrt{k_{\beta_1}^2 - \gamma^2}} \sqrt{\frac{\varepsilon}{2\pi\omega k_1}}, \\ a(\gamma) &= a^{\mathrm{I}}(\gamma)\, \Delta\mu + O(\Delta^2). \end{aligned} \tag{4.202}$$

It is not difficult to show that

$$\operatorname{Re} a^{\mathrm{I}}(\gamma) = O\left(\frac{1}{\gamma^2}\right), \qquad \operatorname{Im} a^{\mathrm{I}}(\gamma) = O\left(\frac{1}{\gamma^3}\right) \tag{4.203}$$

holds for $\gamma \to \infty$, which is completely in agreement with the asymptotic estimation (4.176), and of course $a^{\mathrm{I}}(0) = 0$ holds too. The real and imaginary part of $a^{\mathrm{I}}(\gamma)$ is plotted in Fig. 28. The discussion follows in the next section. Finally, note that the limit for $\Delta\mu \to \infty$, i.e. reflection from a rigid body, is given by $a = -1$ and $a(\gamma) \equiv 0$. This result is obtained without difficulty by modifying the continuity conditions.

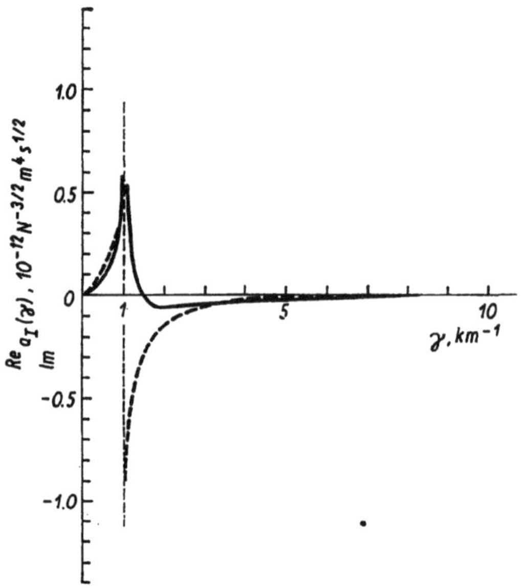

Fig. 28. Real (full) and imaginary (dashed) part of the continuous reflection coefficient in first approximation $a^{\mathrm{I}}(\gamma)$ as a function of the "continuous index" γ

Model parameters: $\mu_1 = 1 \cdot 10^9$ N/m², $\varrho_1 = 1$ g/cm³, $\varepsilon = 1 \cdot 10^6$ N/m³, $\omega = 1$ s⁻¹

4.4.6. *Discussion of the approximate solution*

4.4.6.1. Consideration of the properties of $a^{\mathrm{I}}(\gamma)$

First of all we state that density jumps have no influence on $a(\gamma)$ in the first approximation, which is expressed by the vanishing of $\bar{a}^{\mathrm{I}}(\gamma)$ according to (4.198). The striking property of the first approximation according to (4.202) is the existence of a singularity for $\gamma = k_{\beta_1}$ (see Fig. 28). It follows from equation (3.67) that $\gamma = k_{\beta_1}$ means, in physical terms, perpendicular incidence (concerning the surface) of the corresponding body waves or radiation modes of the continuous spectrum. Here the applied approach fails. This is not surprising, because the same behaviour has already been found, thoroughly discussed and confirmed by an independent numerical analysis already at the closed wave-guide (see sections 4.3.4.1., 4.3.4.2., and 4.3.6.). What has been said is valid to a very large extent for the open wave-guide too, especially the impossibility of expanding (4.186) at this point. In the literature, the existence of such or similar singularities was repeatedly referred to, or these ones can be at least derived from the given formulas. The latter is the case for example in the paper of MARCUSE (1970). In an other context, TIEN et al. (1975) explicitly referred to these problems in considering a film-to-fiber coupler, where the theoretically derived coupling constant of radiation modes is peaked for grazing incidence. Of course, this peak is unphysical and contradicts the observations. To explain this phenomenon it is pointed out in this paper that the wave number of the wave-guide mode should be a complex number instead of a real number used in that calculation, because power is converted from the wave-guide mode to radiation modes. Up to a certain degree this may be correct for that configuration, but not for our model. A similar effect was found for surface waves, incident upon a discontinuity which is formed by a jump of the surface impedance [see Fig. 87 and 88 at NEFEDOV (1979)]. In the paper of RULF and KEDEM (1976) mentioned above, the approximate coefficients of the continuous spectrum were determined in a similar way as in section 4.4.5., and the same singularities occurred. But they were regarded as uncritical, because they are weak enough to ensure the convergence of the corresponding integrals such as $\int_0^\infty a(\gamma)\, U^{(1)}(\gamma)\, \mathrm{d}\gamma$; an important factor.

Finally, the same effect was found with the modified integral equation method according to GELIN et al. (1981), by which the statements of RULF and KEDEM (1976) were confirmed. Incidently, the same singularity occurs in approximating $b(\gamma)$ as well, as can be seen from (4.171).

4.4.6.2. Determination of an approximation of second order for the reflection coefficient a using $a^{\mathrm{I}}(\gamma)$

We conclude from equation (4.169) that a behaves like

$$a = \frac{\mu_1 k_1 - \mu_2 k_2}{\mu_1 k_1 + \mu_2 k_2} + O[(\mu_1 - \mu_2)^2] \tag{4.204}$$

for $\mu_2 \to \mu_1$. For comparison let us give the analogous expression for the transmission coefficient b, too,

$$b = \frac{4\mu_1^2 k_1}{(\mu_1 + \mu_2)(\mu_1 k_1 + \mu_2 k_2)} \cdot \frac{{}^0C^{(1)}}{{}^0C^{(2)}} + O[(\mu_1 - \mu_2)^2] \quad \text{for} \quad \mu_2 \to \mu_1, \tag{4.205}$$

which follows from (4.170). Note that if identifying k with the wave number of SH-waves for a moment, the quantity a in the first approximation coincides with the reflection coefficient of homogeneous SH-waves. In the following, the quadratic term in $\Delta\mu$ is evaluated for the reflection coefficent a. To do this we expand the second term in (4.169) up to the second order in $\Delta\mu$ by using the first approximation of $a(\gamma)$ according to (4.202), and it follows that

$$a = \frac{\mu_1 k_1 - \mu_2 k_2}{\mu_1 k_1 + \mu_2 k_2} + (\Delta\mu)^2 \frac{\varepsilon^3 \mu_1}{\pi} \times \int_0^\infty \frac{\gamma^2 \left(k_1 - \sqrt{k_{\beta_1}^2 - \gamma^2}\right)\left(1 + \frac{1}{k_1}\sqrt{k_{\beta_1}^2 - \gamma^2}\right)}{(\varepsilon^2 + \gamma^2 \mu_1^2)^3 \sqrt{k_{\beta_1}^2 - \gamma^2}} \, \mathrm{d}\gamma + O(\Delta^3), \tag{4.206}$$

which can be simplified as

$$a = \frac{\mu_1 k_1 - \mu_2 k_2}{\mu_1 k_1 + \mu_2 k_2} + (\Delta\mu)^2 \frac{\varepsilon^3}{\pi k_1 \mu_1^5} \times \int_0^\infty \frac{\gamma^2 \, \mathrm{d}\gamma}{\left(\gamma^2 + \frac{\varepsilon^2}{\mu_1^2}\right)^2 \sqrt{k_{\beta_1}^2 - \gamma^2}} + O(\Delta^3). \tag{4.207}$$

By applying the substitution

$$\gamma^2 = \frac{k_{\beta_1}^2 t^2}{1 + t^2}, \qquad \mathrm{d}\gamma = \frac{k_{\beta_1}}{(1 + t^2)^{3/2}} \, \mathrm{d}t \tag{4.208}$$

the integral $\tilde{I}$,

$$\tilde{I} = \int \frac{\gamma^2 \, \mathrm{d}\gamma}{\left(\gamma^2 + \frac{\varepsilon^2}{\mu_1^2}\right) \sqrt{k_{\beta_1}^2 - \gamma^2}}, \tag{4.209}$$

an be transformed into the rational integral

$$\tilde{I} = \int \frac{k_{\beta_1}^2 t^2}{\left(k_1^2 t^2 + \frac{\varepsilon^2}{\mu_1^2}\right)^2} \, \mathrm{d}t \tag{4.210}$$

having the primitive function

$$\tilde{I} = -\frac{t k_{\beta_1}^2}{2 k_1^2 \left(t^2 k_1^2 + \frac{\varepsilon^2}{\mu_1^2}\right)} + \frac{\mu_1 k_{\beta_1}^2}{2 \varepsilon k_1^3} \arctan \frac{t \mu_1 k_1}{\varepsilon}. \tag{4.211}$$

The definite integral I needed in (4.207) is an improper integral (due to the singularity in the denominator as well as to the indefinite integration interval), which has to be evaluated carefully. Using the integration variable t it can be written as

$$I = \int_0^\infty \frac{\gamma^2 \, \mathrm{d}\gamma}{\left(\gamma^2 + \frac{\varepsilon^2}{\mu_1^2}\right) \sqrt{k_{\beta_1}^2 - \gamma^2}} = \int_0^{-\mathrm{i}} \frac{k_{\beta_1}^2 t^2}{\left(k_1^2 t^2 + \frac{\varepsilon^2}{\mu_1^2}\right)} \, \mathrm{d}t, \tag{4.212}$$

and then by using (4.210), (4.211), and the identity

$$\operatorname{arc\,tan}(-\mathrm{i}s) = \frac{\pi}{2} + \frac{\mathrm{i}}{2} \log \frac{s-1}{s+1} \quad \text{for} \quad s > 1 \tag{4.213}$$

the integral is evaluated as

$$I = \frac{\mu_1 k_{\beta_1}^2 \pi}{4 \varepsilon k_1^3} + \mathrm{i} \left\{ \frac{k_{\beta_1}^2}{2k_1^2 \left(\frac{\varepsilon^2}{\mu_1^2} - k_1^2 \right)} + \frac{\mu_1 k_{\beta_1}^2}{2 \varepsilon k_1^3} \log \frac{\mu_1 k_1 - \varepsilon}{\mu_1 k_1 + \varepsilon} \right\}, \tag{4.214}$$

so that finally (4.207) becomes

$$a = \frac{\mu_1 k_1 - \mu_2 k_2}{\mu_1 k_1 + \mu_2 k_2} + (\Delta\mu)^2 \left[\frac{\varepsilon^2 k_{\beta_1}^2}{4 k_1^4 \mu_1^4} + \mathrm{i} \frac{\varepsilon^3}{\pi k_1 \mu_1^5} \times \left\{ \frac{k_{\beta_1}^2}{2k_1^2 \left(\frac{\varepsilon^2}{\mu_1^2} - k_1^2 \right)} + \frac{\mu_1 k_{\beta_1}^2}{2 \varepsilon k_1^3} \log \frac{\mu_1 k_1 - \varepsilon}{\mu_1 k_1 + \varepsilon} \right\} \right] + O(\Delta^3). \tag{4.215}$$

4.4.6.3. The use of $a(\gamma)$ for the determination of the radiation pattern

In this section, it is to be demonstrated how the radiation pattern for the reflected body waves can be obtained from the continuous expansion coefficient of reflected radiation modes, whose first approximation was given under (4.202). Since it is not at all trivial to build up a field of outgoing body waves using the improper eigenfunctions (radiation modes) we will explain it in some detail. The starting point is the integral term U_{rad} of the resulting displacement component $u_{\mathrm{res}}^{(1)}$ according to (4.135),

$$U_{\mathrm{rad}} = \int_0^\infty a(\gamma) \, U^{(1)}(x_3, \gamma) \, \mathrm{e}^{-\mathrm{i}\sqrt{k_{\beta_1}^2 - \gamma^2} x_1} \, \mathrm{d}\gamma, \tag{4.216}$$

which can be written simply as

$$U_{\text{rad}} = \int_{-\infty}^{\infty} a(\gamma)\, e^{i\left(\gamma x_3 - \sqrt{k_{\beta_1}^2 - \gamma^2}\, x_1\right)}\, d\gamma \tag{4.217}$$

by using the expression (3.64) for $U^{(1)}(x_3, \gamma)$ and by considering the relations

$$a(-\gamma) = a^*(\gamma) \quad \text{and} \quad \frac{a(-\gamma)}{a^*(-\gamma)} = -\frac{\varepsilon + i\gamma\mu_1}{\varepsilon - i\gamma\mu_1}, \tag{4.218}$$

which follow immediately from the approximation (4.202). The integral (4.217) can be evaluated approximately using the so-called *saddle-point method.* To do this we have to introduce polar co-ordinates and to transform the integration variable, i.e.

$$\begin{aligned} x_1 &= -\varrho \cos \Theta, \qquad x_3 = \varrho \sin \Theta; \\ \gamma &= k_{\beta_1} \sin \vartheta, \qquad 0 \leqq \Theta \leqq \pi/2. \end{aligned} \tag{4.219}$$

In this way, the integral is brought into the "canonical form"

$$U_{\text{rad}} = U_{\text{rad}}(\varrho, \Theta) = \int_{-(\pi/2)+i\infty}^{(\pi/2)-i\infty} a(k_{\beta_1} \sin \vartheta)\, e^{i\varrho k_{\beta_1} \cos(\vartheta - \Theta)} k_{\beta_1} \cos \vartheta\, d\vartheta. \tag{4.220}$$

Denoting the function

$$F(\vartheta) = \frac{\sin \vartheta\, (k_1 - k_{\beta_1} \cos \vartheta)}{(\varepsilon^2 + k_{\beta_1}^2 \mu_1^2 \sin^2 \vartheta)\, (\varepsilon + i k_{\beta_1} \mu_1 \sin \vartheta)} \tag{4.221}$$

by $F(\vartheta)$ and using (4.202), we obtain

$$U_{\text{rad}}(\varrho, \Theta) = 2i\varepsilon k_{\beta_1} \Delta\mu \sqrt{\frac{\varepsilon}{2\pi\omega k_1}} \int_{-(\pi/2)+i\infty}^{(\pi/2)-i\infty} F(\vartheta)\, e^{i\varrho k_{\beta_1} \cos(\vartheta - \Theta)}\, d\vartheta. \tag{4.222}$$

For the evaluation of this integral, the original integration path $\mathscr{C}$ is transformed into the so-called "saddle point path" or "path of the steepest descent" $\mathscr{C}_s$ which is defined by

$$f(\vartheta) = f(\vartheta_s) - s^2, \quad \text{especially} \quad f(\vartheta) = i k_{\beta_1} \cos(\vartheta - \Theta), \tag{4.223}$$

where s is a real parameter and ϑ_s is the saddle point, defined by [see BREKHOVSKIKH (1960)]

$$\frac{df}{d\vartheta} = 0, \quad \text{especially} \quad \vartheta_s = \Theta. \tag{4.224}$$

In our case, the path of the steepest descent $\mathscr{C}_s$ is obtained from (4.223) by forming the real part as

$$\cos(\vartheta' - \Theta) \cosh \vartheta'' = 1, \tag{4.225}$$

where

$$\vartheta = \vartheta' + i\vartheta'' \tag{4.226}$$

holds. The following Fig. 29 shows the original integration path $\mathscr{C}$ as well as the path of the steepest descent $\mathscr{C}_s$ in the complex ϑ-plane.

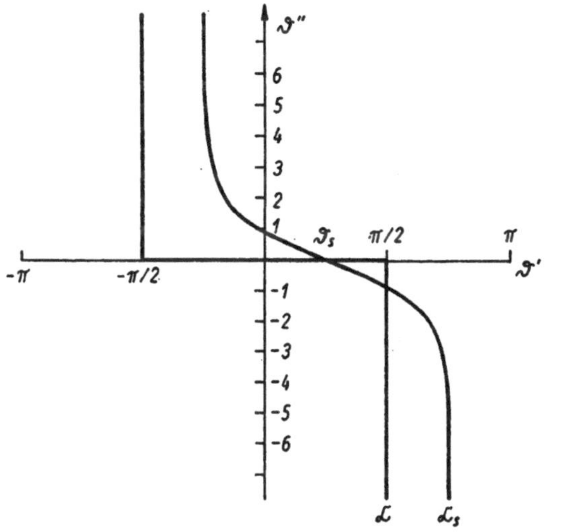

Fig. 29. The original integration path $\mathscr{C}$ and the path of the steepest descent $\mathscr{C}_s$

The deformation of the integration path $\mathscr{C} \to \mathscr{C}_s$ is permissable, because the ends of both contours are situated in regions with $|\vartheta| \to \infty$, where the integrand decreases exponentially according to (4.222). Certainly, at this deformation the contributions of eventually existing branch points and poles have to be taken into account, i.e.

$$U_{\text{rad}} = U^s_{\text{rad}} + U^P, \tag{4.227}$$

where U^s_{rad} means the integral along the path of the steepest descent and U^P summarizes the contributions of the singularities. There are no branch points in our case, but a pole of the first order exists, situated on the positive imaginary axis according to (4.221):

$$\vartheta_P = \mathrm{i}\ \mathrm{arsinh}\ \frac{\varepsilon}{k_{\beta_1}\mu_1}. \tag{4.228}$$

The pole must be considered if it is situated below the integration path. Using (4.225) and (4.228) this is identical with

$$\Theta > \Theta_c, \quad \Theta_c = \arccos \frac{k_\beta}{k_1}, \quad \text{i.e.} \quad k_\beta > k_1 \cos \Theta. \tag{4.229}$$

Accordingly, equation (4.227) has to be specified using Heaviside's function [see (4.19)] in the manner

$$U_{\mathrm{rad}}(\varrho, \Theta) = U_{\mathrm{rad}}^s(\varrho, \Theta) + H(\Theta - \Theta_c)\, U^P(\varrho, \Theta), \tag{4.230}$$

where U^P is given by the expression

$$U^P(\varrho, \Theta) = 2\mathrm{i}\varepsilon k_{\beta_1} \Delta\mu \sqrt{\frac{\varepsilon}{2\pi\omega k_1}}\ 2\pi\mathrm{i}\ \mathrm{Res}\left[F(\vartheta)\ \mathrm{e}^{\mathrm{i}\varrho k_{\beta_1}\cos(\vartheta - \Theta)}\right]_{\vartheta = \vartheta_P} \tag{4.231}$$

and can be determined as follows:

$$\begin{aligned} U^P(\varrho, \Theta) &= -\frac{\varepsilon^2 \Delta\mu}{k_1^2 \mu_1^4} \sqrt{\frac{2\pi\varepsilon}{\omega k_1}}\ \mathrm{e}^{\mathrm{i}\varrho k_{\beta_1}\cos(\vartheta_P - \Theta)} \\ &= -\frac{\varepsilon^2 \Delta\mu}{k_1^2 \mu_1^4} \sqrt{\frac{2\pi\varepsilon}{\omega k_1}}\ \mathrm{e}^{\mathrm{i}\varrho k_1 \cos\Theta}\ \mathrm{e}^{-\frac{\varepsilon}{\mu_1}\varrho \sin\Theta}. \end{aligned} \tag{4.232}$$

Obviously, it is an inhomogeneous wave damped in ϱ-direction. However, it can only be observed in the angular domain $\Theta > \Theta_c$, while the contribution from the saddle point appears in the whole quarter-space. This is a typical situation for such problems. Let us evaluate now U_{rad}^s based on the statements of Brekhovskikh (1960) and Felsen and Marcuvitz (1973). According to the saddle point method an integral of the type

$$I = \int_{\mathscr{C}} \overline{F}(\vartheta)\ \mathrm{e}^{\varrho f(\vartheta)}\ \mathrm{d}\vartheta \tag{4.233}$$

has the following asymptotic expansion

$$I = \mathrm{e}^{\varrho f(\vartheta_s)} \sqrt{\frac{\pi}{\varrho}} \left[\Phi(0) + \frac{1}{4\varrho}\Phi''(0) + \dots\right] \quad \text{for} \quad \varrho \to \infty \tag{4.234}$$

with

$$\Phi(0) = \sqrt{-\frac{2}{f''(\vartheta_s)}} \cdot \overline{F}(\vartheta_s) \tag{4.235}$$

and especially for $f(\vartheta) = i\varphi(\vartheta)$ it is

$$\Phi(0) = \sqrt{\frac{2}{|\varphi''(\vartheta_s)|}} \cdot \overline{F}(\vartheta_s)\, e^{i\frac{\pi}{4}\operatorname{sgn}\varphi''(\vartheta_s)}. \tag{4.236}$$

Accordingly, we obtain the following asymptotic expression for the contribution of the saddle point U^s_{rad} for great ϱ:

$$U^s_{\text{rad}} = 2i\varepsilon k_{\beta_1}\Delta\mu \sqrt{\frac{\varepsilon}{2\pi\omega k_1}} \int_{\mathscr{C}_s} F(\vartheta)\, e^{i\varrho k_{\beta_1}\cos(\vartheta-\Theta)}\, d\vartheta$$

$$= 2i\varepsilon\Delta\mu \sqrt{\frac{\varepsilon k_{\beta_1}}{\omega}}\, F(\Theta)\, \frac{e^{i\left(\varrho k_{\beta_1}-\frac{\pi}{4}\right)}}{\sqrt{\varrho k_1}} \left[1 + O\left(\frac{1}{\varrho k_{\beta_1}}\right)\right]$$

$$\text{for} \quad \Theta \neq 0 \quad \text{and} \quad \varrho \to \infty. \tag{4.237}$$

For $\Theta = 0$ the function $F(\Theta)$ vanishes and we have to consider a further term in the expansion (4.234). Especially for $\overline{F}(\vartheta_s) = 0$ it holds that

$$\frac{1}{2}\, \Phi''(0) = -\sqrt{-\frac{2}{f''(\vartheta_s)}}\, \frac{\overline{F}''(\vartheta_s)}{f''(\vartheta_s)}, \tag{4.238}$$

and by using the expression

$$F''(0) = -\frac{2ik_{\beta_1}\mu_1(k_1 - k_{\beta_1})}{\varepsilon^4} \tag{4.239}$$

in our case we obtain the estimation

$$U^s_{\text{rad}}(\varrho, 0) = -2i\mu_1\Delta\mu k_1 \sqrt{\frac{\varepsilon k_{\beta_1}}{\omega}}\, \frac{(k_1 - k_{\beta_1})}{\varepsilon^3}\, \frac{e^{i\left(\varrho k_{\beta_1}-\frac{\pi}{4}\right)}}{(\varrho k_1)^{3/2}}$$

$$\times \left[1 + O\left(\frac{1}{\varrho^2 k^2_{\beta_1}}\right)\right] \quad \text{for} \quad \varrho \to \infty. \tag{4.240}$$

We see from (4.237) and (4.240) that U^s_{rad} behaves like an outgoing cylindrical wave at great distance from the discontinuity and is dominating in comparison with the contribution of the pole, because the latter decreases exponentially with the distance according to (4.232).

Let us evaluate now the so-called power radiation pattern of this cylindrical wave. The time-averaged ϱ-component of the energy flux is given by the expression [compare with (3.28) and Tuan and Chang (1972)]

$$S_\varrho = \frac{i\omega}{4}\, \mu \left(U \frac{\partial U^*}{\partial \varrho} - U^* \frac{\partial U}{\partial \varrho}\right). \tag{4.241}$$

Identifying U with U^s_{rad} according to (4.237), we obtain

$$S_\varrho = \frac{1}{\varrho}\,\hat{F}(\Theta), \tag{4.242}$$

where $\hat{F}(\Theta)$ is the dimensionless *power radiation pattern*, defined by

$$\hat{F}(\Theta) = \frac{k_{\beta_1}}{k_1}\cdot 2\mu_1 k_{\beta_1}\varepsilon^3 \Delta\mu^2 \cdot \frac{\sin^2\Theta(k_1 - k_{\beta_1}\cos\Theta)^2}{(\varepsilon^2 + k^2_{\beta_1}\mu_1^2 \sin^2\Theta)^3}. \tag{4.243}$$

This expression contains implicitly the fact that the incident surface wave transports unit energy. The radiation pattern $\hat{F}(\Theta)$ is plotted in Fig. 30 for a simple model for two ε-values.

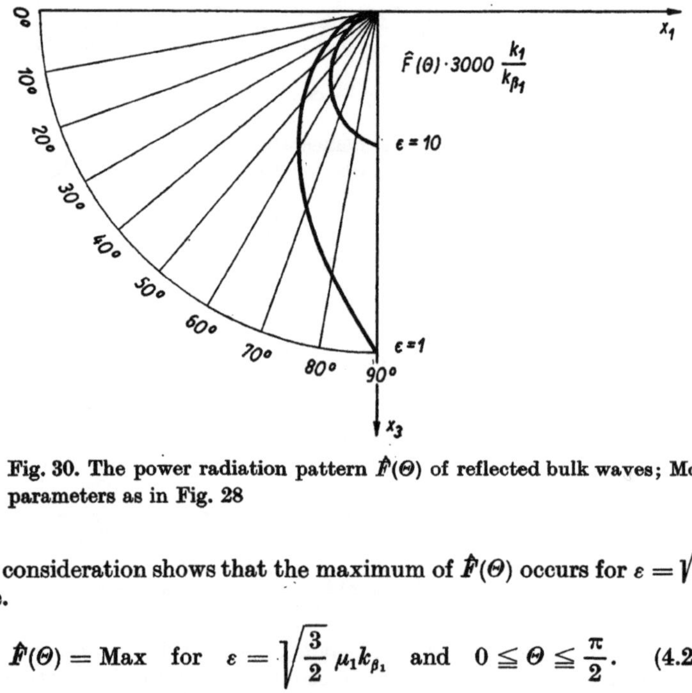

Fig. 30. The power radiation pattern $\hat{F}(\Theta)$ of reflected bulk waves; Model parameters as in Fig. 28

A simple consideration shows that the maximum of $\hat{F}(\Theta)$ occurs for $\varepsilon = \sqrt{3/2} \times \mu_1 k_{\beta_1}$, i.e.

$$\hat{F}(\Theta) = \text{Max} \quad \text{for} \quad \varepsilon = \sqrt{\frac{3}{2}}\,\mu_1 k_{\beta_1} \quad \text{and} \quad 0 \leqq \Theta \leqq \frac{\pi}{2}. \tag{4.244}$$

The total reflected bulk wave power (i.e. the contribution from the saddle point) converted at the discontinuity is given by the expression

$$S^s_{rad} = \int\limits_0^{\pi/2} S_\varrho(\varrho, \Theta)\,\varrho\,d\Theta = \int\limits_0^{\pi/2} \hat{F}(\Theta)\,d\Theta. \tag{4.245}$$

Analogous considerations can be carried out for the transmitted energy.

4.4.7. *The exact solution according to Wiener-Hopf for the impedance jump discontinuity*

In the following, an analytical solution for a discontinuity of the impedance jump type (see Fig. 27a) is given on the basis of the *Wiener-Hopf technique.* Since the derivation is very extensive and the train of thought should not be disturbed we will summarize here the final results only, while the details will follow in chapter 5. The complete expression for the reflection coefficient R (being equivalent with a) of an incident surface wave at an impedance jump ($\Delta\varepsilon \neq 0$, $\Delta\mu = 0$, $\Delta\varrho = 0$) is

$$R = \frac{k_1 - k_2}{k_1} \cdot \frac{\varepsilon_1}{\varepsilon_1 + \varepsilon_2} \, e^{i\varphi_R}, \tag{4.246}$$

with

$$\varphi_R = \frac{1}{\pi} (E_1 - E_2), \quad E_1 = \int\limits_{\xi_1 - \xi_2}^{\xi_1 + \xi_2} \frac{s \, ds}{\sinh s}, \quad E_2 = \int\limits_0^{2\xi_1} \frac{s \, ds}{\sinh s} \tag{4.247}$$

and

$$\xi_1 = \operatorname{arcosh} \frac{k_1}{k_\beta}, \quad \xi_2 = \operatorname{arcosh} \frac{k_2}{k_\beta}. \tag{4.248}$$

Since $\Delta\mu = 0$ and $\Delta\varrho = 0$ the notation $k_\beta = k_{\beta_1} = k_{\beta_2}$ has been introduced. Further on, the transmission coefficient T_r of the surface waves is given by

$$T_r = \frac{2\sqrt{k_1 k_2}}{k_2} \cdot \frac{\varepsilon_2}{\varepsilon_1 + \varepsilon_2} \, e^{i\varphi_T} \tag{4.249}$$

with

$$\varphi_T = \frac{1}{2\pi} (E_1 - E_2 - E_3 + E_4),$$
$$E_3 = \int\limits_0^{2\xi_2} \frac{s \, ds}{\operatorname{arsinh} s}, \qquad E_4 = \int\limits_{\xi_2 - \xi_1}^{\xi_2 + \xi_1} \frac{s \, ds}{\operatorname{arsinh} s}. \tag{4.250}$$

This coefficient is connected with b according to

$$T_r = \frac{{}^0C^{(2)}}{{}^0C^{(1)}} \, b \tag{4.251}$$

and it expresses the amplitude ratio at the surface. The absolute values of the reflection and transmission coefficients have the simple form

$$|R| = \frac{|k_1 - k_2|}{k_1} \cdot \frac{\varepsilon_1}{\varepsilon_1 + \varepsilon_2} \tag{4.252}$$

and

$$|T_r| = \frac{2\sqrt{k_1 k_2}}{k_2} \cdot \frac{\varepsilon_2}{\varepsilon_1 + \varepsilon_2}. \tag{4.253}$$

These formulas will be the starting point for the comparison with various approximate solutions which do not take into account the imaginary parts in any case. An analysis of the phase factors exp $(i\varphi_R)$ and exp $(i\varphi_T)$ will still be carried out in connection with a model calculation (see section 4.4.9.4.2.).

4.4.8. *Classification of the discontinuities under consideration*

4.4.8.1. Condition for the existence of an elementary solution of the discontinuity problem

It is convenient to introduce the following dimensionless parameters $\varkappa_1$, $\varkappa_2$, and $\varkappa_3$, defined by

$$\varepsilon_2 = \varkappa_1 \varepsilon_1, \qquad \mu_2 = \varkappa_2 \mu_1, \qquad \varrho_2 = \varkappa_3 \varrho_1, \tag{4.254}$$

to characterize the media right and left of the discontinuity. For the sake of simplicity we will set later on $\varepsilon_1 \overset{!}{=} \varepsilon$, $\mu_1 \overset{!}{=} \mu$, $\varrho_1 \overset{!}{=} \varrho$.

Let us now consider the question of how the discrete eigenfunctions, i.e. the surface waves, have to be constituted right and left of the discontinuity so that the continuity conditions can be fulfilled alone with them. In fact, it is the different damping behaviour with depth of the discrete eigenfunctions in the media 1 and 2 which is responsible for the conversion into body waves or radiation modes, and for the difficulties connected with it. The damping behaviour is controlled by the exponent in the e-function [see (3.63)] and consequently

$$\frac{\varepsilon_1}{\mu_1} = \frac{\varepsilon_2}{\mu_2} \tag{4.255}$$

is a necessary condition for the existence of an elementary solution, which can be simply written as

$$\varkappa_1 = \varkappa_2 \tag{4.256}$$

using (4.254). By solving the problem for this case, we will show that this condition is also sufficient. But first, let us demonstrate how it is possible to obtain this condition in another way. Proceeding from the modified equations (4.149)—(4.152) for the combination "impedance jump + vertical discontinuity", the reflection coefficient a is obtained in complete analogy to (4.153) as

$$a = \hat{a} - \int_0^\infty \ldots \mathrm{d}\gamma \tag{4.257}$$

and using a modified quantity P according to (4.164) the transmission coefficient b is given by

$$b = \hat{b} + \int_0^\infty \ldots \mathrm{d}\gamma, \tag{4.258}$$

where

$$\hat{a} = \frac{\mu_1 k_1 - \mu_2 k_2}{\mu_1 k_1 + \mu_2 k_2} \tag{4.259}$$

and

$$\hat{b} = \frac{4\mu_1^2 k_1}{\mu_1 k_1 + \mu_2 k_2} \cdot \frac{\varepsilon_2}{\varepsilon_1 \mu_2 + \varepsilon_2 \mu_1} \cdot \frac{{}^0C^{(1)}}{{}^0C^{(2)}}. \tag{4.260}$$

The occurring integrals are proportionals to $Q(\gamma)$ and vanish according to (4.165) for

$$\varepsilon_1 \Delta\mu = \mu_1 \Delta\varepsilon, \tag{4.261}$$

being identical with (4.255) or (4.256). For this special case, the reflection and transmission coefficients of surface waves have the simple form

$$\hat{a}^0 = \hat{a} \qquad \hat{b}^0 = \frac{2\mu_1 k_1}{(\mu_1 k_1 + \mu_2 k_2)} \cdot \frac{{}^0C^{(1)}}{{}^0C^{(2)}}. \tag{4.262}$$

It is easy to recognize that the system "incident — reflected — transmitted surface wave" fulfils the continuity conditions at the boundary exactly, so long as the condition (4.261) holds. It follows that the scalar products $Q(\gamma)$ and $R(\gamma)$ vanish according to (4.165) and (4.166), so that the integral equation (4.154) for $a(\gamma)$ is reduced to

$$\int_0^\infty a(\sigma) \left\{ \left(\sqrt{k_{\beta_2}^2 - \gamma^2} + \frac{\mu_1}{\mu_2} \sqrt{k_{\beta_1}^2 - \sigma^2} \right) T(\sigma, \gamma) \right\} \mathrm{d}\sigma = 0 \tag{4.263}$$

and using (4.167) we obtain immediately

$$a(\gamma) \equiv 0 \tag{4.264}$$

and similarly

$$b(\gamma) \equiv 0. \tag{4.265}$$

Consequently, by combining the jumps $\Delta\mu$ and $\Delta\varepsilon$ in a suitable manner, the body waves or radiation modes can be completely suppressed. This compensation effect has found numerous technical applications. An analogue for seismic surface waves is the so-called Satô model [Satô (1961)], which consists of a layer over the half-space, where the elastic parameters left and right of the vertical discontinuity have to fulfil certain conditions. These conditions originate from Higuchi (1932) and are given by (see Fig. 31)

$$\frac{\mu_1^{(1)}}{\mu_2^{(1)}} = \frac{\mu_1^{(2)}}{\mu_2^{(2)}}, \qquad \frac{1}{\beta_1^{(1)^2}} - \frac{1}{\beta_1^{(2)^2}} = \frac{1}{\beta_2^{(1)^2}} - \frac{1}{\beta_2^{(2)^2}}. \tag{4.266}$$

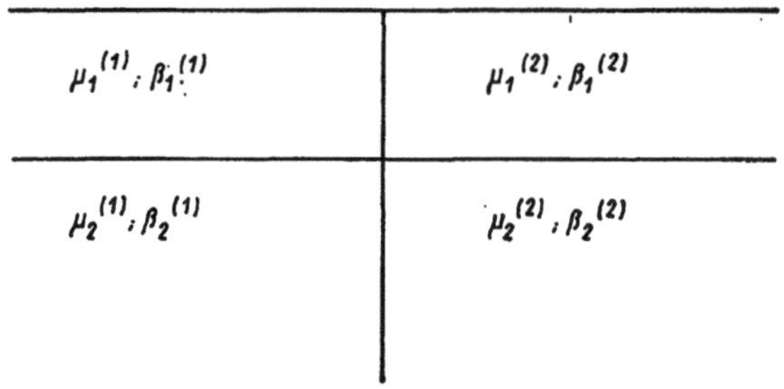

Fig. 31. Geometry of the Satô model

They enable the reflection and transmission coefficients of Love waves for vertical incidence to be determined in a simple manner analogously to (4.262). Recently, the conditions (4.266) were generalized by Kazi and Niazy (1982b), first for the 2-layer case and then for the n-layer case as well.

4.4.8.2. Condition for the transition impedance jump — vertical discontinuity

First, let us give the condition for the existence of an impedance jump using the parameters defined under (4.254),

$$\varkappa_2 = \varkappa_3 = 1, \qquad \varkappa_1 \neq 1. \tag{4.267}$$

Since this problem has an exact solution according to Wiener-Hopf the question arises of whether it is possible to transfer these results to a jump of material parameters μ and ϱ, characterized by

$$\varkappa_1 = 1, \qquad \varkappa_2 \neq 1, \qquad \varkappa_3 \neq 1. \tag{4.268}$$

First, these two cases have to coincide as regards the damping behaviour with depth of the discrete eigenfunctions. Using (3.63) this fact leads to the condition

$$\varkappa_1 \cdot \varkappa_2 = 1. \tag{4.269}$$

Further on, the application of the Wiener-Hopf method is connected with a unique equation of motion

$$\Delta u + k_\beta^2 u = 0, \qquad u_2(x_1, x_3, t) \overset{!}{=} u(x_1, x_3)\, \mathrm{e}^{-\mathrm{i}\omega t} \tag{4.270}$$

for both regions right and left of the discontinuity. So it is necessary to require a unique k_β for both kinds of discontinuities and the condition

$$\varkappa_2 = \varkappa_3 \tag{4.271}$$

follows.

Consequently, the results of the impedance jump problem can only be transferred to the vertical discontinuity, if (4.269) and (4.271) hold.

4.4.9. *Application of the LQBR-method to several test cases and comparison with the approximate solution and the exact one*

4.4.9.1. General remarks

For the problems under consideration, analytical solutions are available in a few special cases only. That is why the search for both simple and effective approximation methods is understandable. In mathematical physics, the solution of boundary problems is often accomplished by applying collocation and least-square methods. Here, advantage is taken of the fact that one can easily satisfy the differential equations of the problem and the latter one is reduced to satisfying boundary conditions over certain interfaces. According to Davies (1973) a boundary residual $\mathcal{R}(s)$ is defined as a linear combination of the total field expansions in adjacent regions such that the vanishing of $\mathcal{R}(s)$ is a necessary (and preferably sufficient) condition for the unique physical solution to the problem. Since we are only able to take into account a finite number of terms from the field expansions, we are looking for methods to satisfy

$$\mathcal{R}(s) = 0 \tag{4.272}$$

approximately. The simplest method is undoubtedly point matching (or collocation method), where $\mathcal{R}(s)$ is made to vanish at a finite number of selected points:

$$\mathcal{R}(s_1) = \mathcal{R}(s_2) = \cdots = \mathcal{R}(s_n) = 0. \tag{4.273}$$

A fundamental difficulty with this method is the tacit assumption of convergence with an increasing number of points.

Truncated series at both sides of the discontinuity are the basis of the "*Least-square boundary residual*" method (LQBR-method), where a convenient error functional F, symbolically written as

$$F = F(\mathcal{R}), \tag{4.274}$$

is minimized on the interface:

$$F \stackrel{!}{=} \text{Min}. \tag{4.275}$$

The minimization is with respect to the same linear parameters, i.e. the expansion coefficients, just as in the collocation method and the latter are determined in this way. The stationarity problem for F is equivalent to the original boundary value problem. The success of this method depends both on the suitable choice of the error functional and the basis functions. Precisely there is a considerable freedom, which has already led to difficulties as we will see later on. That is why the choice of the error functional should be carried out very carefully. On the other hand, Davies (1973) was able to prove that the least-square method is a rigorously convergent procedure, i.e. if the number of terms is increased in the various expansions, the resulting approximate solutions will approach the physically correct fields. Also Millar (1973) has proved for a whole class of scattering problems, that the representation which follows from the minimization of the error functional, converges uniformly to the correct solution of the problem, provided that a complete set of basis functions is chosen. In recent years, Herrera and co-workers [see e.g. Herrera (1977), Herrera (1979), England et al. (1980)] have developed a *theory of connectivity* which allows a systematic formulation of boundary value problems and which naturally leads to complete eigenfunction systems. At the same time, the least-square method obtains a deep mathematical foundation.

4.4.9.2. Derivation of the LQBR-method for the discontinuity models under consideration

Let us begin with some remarks concerning the eigenfunction system. If we had a homogeneous half-space with a disturbance at the surface, then those eigenfunctions would be adequate, which solve the boundary value problem for the unmodified half-space [see e.g. England et al. (1980)]. In our case, the discrete and continuous modes of media 1 and 2 are adequate. As is known, they form a complete system, but its handling is troublesome. That is the reason why several authors use other complete eigenfunction systems, such as Laguerre polynomials [see e.g. Mahmoud and Beal (1975) or Rozzi (1978)], Hankel functions [see England et al. (1980)] or Hermite-Gaussian functions [see Pudensi and Ferreira (1982)]. Certainly, in this way the basis functions are discretized, but the physical clearness has been partly lost. A compromise would be an eigenfunction system consisting of the discrete surface wave modes complemented by Laguerre polynomials to cover the continuous spectrum represented by the body waves [see Rozzi (1978)]. Here, we cannot follow this approach further. On the contrary, the LQBR-method is derived on the basis of discrete surface waves alone (for the homogeneous half-space with an impedance surface on the basis of only one surface wave mode). Obviously, this eigenfunction system is not complete for the open wave-guide and thus a convergence in the sense mentioned above cannot be achieved in this way. On the other hand, this method has found a considerable application in seismology (see section 4.5.3.2.1.) and therefore it is necessary to examine its relevance critically.

After having dealt with the question about the eigenfunction system let us speak about the error functional. The continuity conditions (4.144) are written as

$$|\psi_1^{(+)}\rangle + a\,|\psi_1^{(-)}\rangle = b\,|\psi_2^{(+)}\rangle + |\chi\rangle, \tag{4.276}$$

where $|\chi\rangle$ is a residual vector,

$$|\chi\rangle = \begin{pmatrix} \mathrm{i}\omega U \\ S_{12} \end{pmatrix}, \tag{4.277}$$

which contains the displacement and stress discontinuities caused by the impossibility of satisfying the continuity conditions on the basis of surface waves alone. Consequently, the vector $|\chi\rangle$ contains symbolically the two integral terms of (4.144). Fully written, (4.276) reads:

$$U_1 + aU_1 = bU_2 + U, \tag{4.278}$$

$$S_{12}^{(+)(1)} + aS_{12}^{(-)(1)} = bS_{12}^{(+)(2)} + S_{12}. \tag{4.279}$$

A functional F_1, defined by

$$F_1 = \int_0^\infty S_{12}S_{12}^* \,\mathrm{d}x_3, \tag{4.280}$$

was introduced by ALSOP (1966) in a more or less arbitrary manner. It indicates the mean square of the stress discontinuity over the interface. This functional is a function of a and b, but by applying the orthonormality relation (2.35), which can be also written as [see MALISCHEWSKY (1973)]

$$\frac{\mathrm{i}\omega}{4}\int_0^\infty ({}^\alpha U {}^\beta S_{12}^* - {}^\beta U^* {}^\alpha S_{12})\,\mathrm{d}x_3 = {}^\alpha C\delta_{\alpha\beta}, \tag{4.281}$$

to (4.278) and (4.279) and by neglecting U and S_{12}, a linear relation between a and b is obtained, so that F_1 can be expressed for example as a function of a alone. Then the minimum condition for F_1 yields the approximated reflection coefficient a_{Al} and by substituting into the linear relation between a and b the approximated transmission coefficient b_{Al} is obtained. The insufficiency of this procedure cannot be discussed here, but we shall give the final results for comparison. First, the quantities P_{12}, P_{21}, $\tilde{P}$, $\tilde{Q}$, defined by

$$P_{12} = \int_0^\infty U_1 S_{12}^{(2)*}\, \mathrm{d}x_3 = -\mathrm{i}\mu_2 k_2 P, \tag{4.282}$$

$$P_{21} = \int_0^\infty U_2^* S_{12}^{(1)}\, \mathrm{d}x_3 = \mathrm{i}\mu_1 k_1 P,$$

$$\tilde{P} = \frac{\mathrm{i}\omega}{4}(P_{12} - P_{21}), \qquad \tilde{Q} = \frac{\mathrm{i}\omega}{4}(P_{12} + P_{21}) \tag{4.283}$$

are introduced, which will be required later on. By using (4.164) we obtain

$$\tilde{P} = \frac{\omega}{4} \frac{\mu_1\mu_2{}^0C^{(1)0}C^{(2)}}{\varepsilon_1\mu_2 + \varepsilon_2\mu_1}(\mu_2 k_2 + \mu_1 k_1) \tag{4.284}$$

and

$$\tilde{Q} = \frac{\omega}{4} \frac{\mu_1\mu_2{}^0C^{(1)0}C^{(2)}}{\varepsilon_1\mu_2 + \varepsilon_2\mu_1}(\mu_2 k_2 - \mu_1 k_1). \tag{4.285}$$

The linear relation between the reflection and transmission coefficients follows as

$$b = \tilde{P} + a\tilde{Q} - \frac{\mathrm{i}\omega}{4}\int_0^\infty [U S_{12}^{(2)*} - U_2^* S_{12}]\, \mathrm{d}x_3, \tag{4.286}$$

and by neglecting the integral term on the right side Alsop's approximation

$$b_{\mathrm{Al}} = \tilde{P} + a_{\mathrm{Al}}\tilde{Q} \tag{4.287}$$

is obtained. For a vertical discontinuity ($\Delta\mu$, $\Delta\varrho \neq 0$; $\Delta\varepsilon = 0$), the quantity a_{Al} is

$$a_{\mathrm{Al}} = \frac{\mu_1\mu_2 k_1^2 \Delta\mu + \mu_1^3 k_1^2 - \mu_2^3 k_2^2}{\mu_1\mu_2 k_1^2 \Delta\mu + \mu_1^3 k_1^2 + \mu_2^3 k_2^2 + 4\mu_1\mu_2^2 k_1 k_2}. \tag{4.288}$$

It has already been pointed out by MALISCHEWSKY (1976) that a_{Al} and b_{Al} are necessarily real quantities. For the problem under consideration, a functional F_2 being more adequate in every respect is obtained in a natural way by taking the considered vector space as a basis and by assuming

$$F_2 \sim \langle \chi \mid \chi \rangle, \tag{4.289}$$

where the definition (4.8) of the scalar product is used, e.g.

$$F_2 = \int_0^\infty (U S_{12}^* - U^* S_{12})\, \mathrm{d}x_3. \tag{4.290}$$

According to HERRERA (1982) the form of F_2 which is connected with the orthonormality relation (4.281) is a consequence of the *biorthogonality* of the used eigenfunction system. HERRERA (1981) and HERRERA and SPENCE (1981) have described the theory of biorthogonal function systems and its application for the solution of various boundary value problems in a general manner. Appearing for the first time in geophysical literature, HERRERA (1964) applied this theory to the derivation of the orthogonality relation of Rayleigh waves. In the paper of HERRERA and SPENCE (1981), it was pointed out how this procedure can be generalized to obtain biorthogonal systems of solutions for a wide class of equations. The fundamental idea is the consideration of a linear space of solutions N_P of the given partial differential equation, which can be decomposed into two linear subspaces N_P^1 and N_P^2 being commutative as regards an antisymmetric bilinear functional A_0 defining the biorthogonality relation [compare with (4.281]). In our case, these subspaces correspond with those denoted in section 4.2.1. with (+) and (−), where the both families of solutions

$$^{\alpha}U\,\mathrm{e}^{\mathrm{i}^{\alpha}kx_1} \in N_P^1 \quad \text{and} \quad {}^{\alpha}U^*\,\mathrm{e}^{-\mathrm{i}^{\alpha}kx_1} \in N_P^2, \qquad \alpha = 1, 2, \ldots, \tag{4.291}$$

are conjugate complex and necessarily biorthogonal concerning A_0. These limited remarks about the very general theory must suffice here.

According to (4.276) or (4.278) and (4.279) the functional F_2 is a function of a and b,

$$F_2 = F_2(a, b), \tag{4.292}$$

and the stationarity condition for this functional,

$$\frac{\partial F_2}{\partial a} = \frac{\partial F_2}{\partial b} = 0, \tag{4.293}$$

yields the LQBR-approximation of the reflection and transmission coefficients as

$$a_{\mathrm{LQ}} = \frac{-\tilde{P}\tilde{Q}}{\dfrac{\omega^2}{16}\,\dfrac{\mu_1^2\mu_2^2k_1k_2}{\varepsilon_1\varepsilon_2}\,{}^0C^{(1)2}\,{}^0C^{(2)2} + \tilde{Q}^2} \tag{4.294}$$

and

$$b_{\mathrm{LQ}} = \frac{\omega}{4}\,\frac{\mu_1^2k_1}{\varepsilon_1}\,{}^0C^{(1)2}\,\frac{\tilde{P}}{\dfrac{\omega^2}{16}\,\dfrac{\mu_1^2\mu_2^2k_1k_2}{\varepsilon_1\varepsilon_2}\,{}^0C^{(1)2}\,{}^0C^{(2)2} + \tilde{Q}^2}, \tag{4.295}$$

and finally it follows that

$$a_{\mathrm{LQ}} = \frac{\mu_1^2k_1^2 - \mu_2^2k_2^2}{(\mu_2k_2 - \mu_1k_1)^2 + \dfrac{k_1k_2}{\varepsilon_1\varepsilon_2}(\varepsilon_1\mu_2 + \varepsilon_2\mu_1)^2} \tag{4.296}$$

and

$$b_{LQ} = \frac{\mu_1 k_1(\mu_1 k_1 + \mu_2 k_2)}{\varepsilon_1\mu_2(\varepsilon_1\mu_2 + \varepsilon_2\mu_1)\left\{\frac{k_1 k_2}{\varepsilon_1\varepsilon_2} + \left(\frac{\mu_2 k_2 - \mu_1 k_1}{\varepsilon_1\mu_2 + \varepsilon_2\mu_1}\right)^2\right\}} \cdot \frac{{}^0C^{(1)}}{{}^0C^{(2)}}. \tag{4.297}$$

For eigenfunctions transporting unit energy according to (3.76), the equations (4.294) and (4.295) have the simple form

$$a_{LQ} = -\frac{\tilde{P}\tilde{Q}}{1 + \tilde{Q}^2} \quad \text{and} \quad b_{LQ} = \frac{\tilde{P}}{1 + Q^2}. \tag{4.298}$$

Obviously, in the limit $H \to \infty$ the former formulas (4.119) and (4.120) for the closed wave-guide and the special case $\varepsilon_1 = \varepsilon_2$ change into the corresponding formulas (4.296) and (4.297). Analogously to (4.287) it holds that

$$b_{LQ} = \tilde{P} + a_{LQ}\tilde{Q} \tag{4.299}$$

and further on

$$\operatorname{Im} a_{LQ} \equiv 0, \qquad \operatorname{Im} b_{LQ} \equiv 0, \tag{4.300}$$

i.e. possible phase jumps unequal to 0 or π are not included in this approximation.

As can be seen, the transmission coefficient b_{LQ} depends on the normalization of the eigenfunctions left and right of the discontinuity. For the LQBR-approximation, the actually interesting and measurable ratio of the wave fields at the surface $x_3 = 0$ is denoted by B_{LQ} and can be calculated as [see also (4.251)!]

$$B_{LQ} = b_{LQ} \cdot \frac{{}^0C^{(2)}}{{}^0C^{(1)}}. \tag{4.301}$$

Analogously to (4.128), we are able to define again a quantity PEUF,

$$\text{PEUF} = (1 - a^2 - b^2) \cdot 100\%, \tag{4.302}$$

expressing the percentage of energy which, during the interaction with the discontinuity, does not remain in the surface wave field for this approximation. In this connection, let us refer to the following fact. Comparing (3.28) and (4.290), we can identify F_2, up to a factor $i\omega/4$, with that energy flux which is produced by the "residues" U and S_{12} and according to (4.278) (4.279), and (4.290) the explicit expression is given by

$$F_2 = \frac{4i}{\omega}(a^2 - b^2 - 1 + 2b\tilde{P} + 2ab\tilde{Q}). \tag{4.303}$$

Using the relations (4.298) and (4.299) we realize that F_2 for the stationary point, i.e. $F_2(a_{LQ}, b_{LQ})$, adopts the value

$$\frac{i\omega}{4} F_2(a_{LQ}, b_{LQ}) = 1 - a_{LQ}^2 - b_{LQ}^2, \tag{4.304}$$

so that this energy flux is identical with the one produced by the incident, reflected, and transmitted surface waves. That means that for this approximation a certain kind of "energy conservation law" is valid, which has already been demonstrated by LEVSHIN and YANOVSKAYA (1976) and ITS and YANOVSKAYA (1977) for Love and Rayleigh waves respectively. However, this behaviour should not obscure the fact that it is also only an approximation, because the approximated equation (4.299) [and not the exact one (4.286)] is a necessary (and sufficient) condition for the validity of (4.304). The apparent contradiction is solved by considering that equation (4.304) is in no case identical with

$$P_{\text{inc}} = P_{\text{R}} + P_{\text{T}} + P_{\text{rad}} \tag{4.305}$$

characterizing the energy conservation at the discontinuity, where P_{inc} is the incident, P_{R} the reflected, and P_{T} the transmitted energy of the surface waves, and P_{rad} the energy of the radiation modes (body waves). Certainly, F_2 looks like an energy flux but the stress and displacement discontinuities occur at the interface of two different media and the energy radiated into these cannot be obtained on the basis of a common expression. On the contrary, the more complicated formulas like (4.245) and another one for medium 2 have to be applied.

4.4.9.3. Comparison of various least-squares methods

Of course, in addition to F_1 and F_2 it is possible to construct as many other functionals as you like. Similarly to MORISHITA et al. (1979) a functional F_3 could be defined as

$$F_3 = \int_0^\infty [UU^* + \bar{\eta} S_{12} S_{12}^*] \, \mathrm{d}x_3, \tag{4.306}$$

but the parameter $\bar{\eta}$ remains arbitrary. Further possibilities would be the quadratic "energy flux"

$$F_4 = \int_0^\infty [US_{12}^* - U^* S_{12}] \, [U^* S_{12} - US_{12}^*] \, \mathrm{d}x_3, \tag{4.307}$$

or

$$F_5 = \int_0^\infty UU^* S_{12} S_{12}^* \, \mathrm{d}x_3, \tag{4.308}$$

$$F_6 = \int_0^\infty UU^* \, \mathrm{d}x_3 \cdot \int_0^\infty S_{12} S_{12}^* \, \mathrm{d}x_3 \quad \text{and so on.} \tag{4.309}$$

These functionals and further ones were tested by comparing the results with the exact solutions according to Wiener-Hopf. The results were in each case not so accurate as those on the basis of F_2, although they were partly better than Alsop's ones (F_1). This may be an additional verification for the fact that the functional F_2 is optimal for this kind of approximation. Since Alsop's method has nevertheless found a certain application, we want to compare here the quantities a_{AL}, b_{AL}, a_{LQ}, b_{LQ}, $\hat{a}$, and $\hat{b}$ based on a simple test case. Proceeding from a simple vertical discontinuity with $\varkappa_1 = \varkappa_3 = 1$ and using (4.259), (4.288), and (4.298) the following formulas are obtained:

$$a_{AL} = \frac{(1 - \varkappa_2 + \varkappa_2^2)\,k_1^2 - \varkappa_2^3 k_2^2}{(1 - \varkappa_2 + \varkappa_2^2)\,k_1^2 + \varkappa_2^3 k_2^2 + 4\varkappa_2^2 k_1 k_2}, \tag{4.310}$$

$$a_{LQ} = \frac{k_1^2 - \varkappa_2^2 k_2^2}{k_1^2 + \varkappa_2^2 k_2^2 + k_1 k_2 (1 + \varkappa_2^2)}, \tag{4.311}$$

$$\hat{a} = \frac{k_1 - \varkappa_2 k_2}{k_1 + \varkappa_2 k_2}. \tag{4.312}$$

The quantities b_{AL} and b_{LQ} are calculated according to (4.287) and (4.299), and if we normalize for unit energy transport $\hat{b}$ is obtained from (4.260) as

$$\hat{b} = \frac{4\varkappa_2 k_1}{(1 + \varkappa_2)\,(k_1 + \varkappa_2 k_2)} \sqrt{\frac{k_2}{k_1}}. \tag{4.313}$$

For the test case, the same model parameters are used as in Fig. 28. The reflection coefficients a_{AL}, a_{LQ}, and $\hat{a}$ and the transmission coefficients b_{AL}, b_{LQ}, and $\hat{b}$ are given as functions of $\varkappa_2$ in Fig. 32 and Fig. 33, respectively.

Obviously, the curve for a_{LQ} osculates with curve $\hat{a}$ in a wide range, while for a_{AL} this is the case in the range $0.7 < \varkappa_2 < 1.1$ only. The transmission coefficients b_{LQ} and $\hat{b}$ hardly differ from each other in the range under consideration, while the coefficient b_{AL} shows a comparatively small deviation

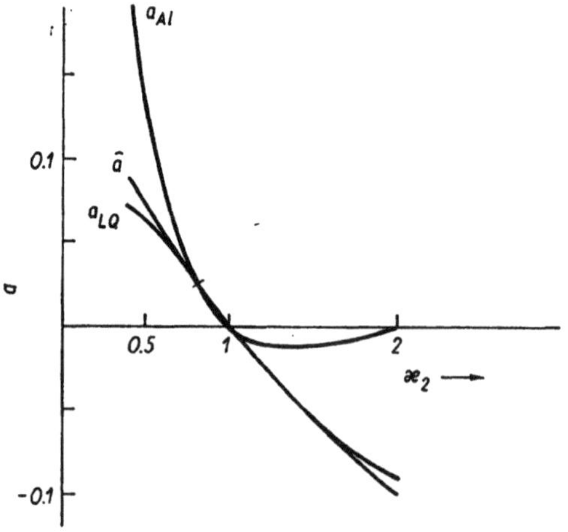

Fig. 32. Comparison of the reflection coefficients a_{Al}, a_{LQ}, and $\hat{a}$

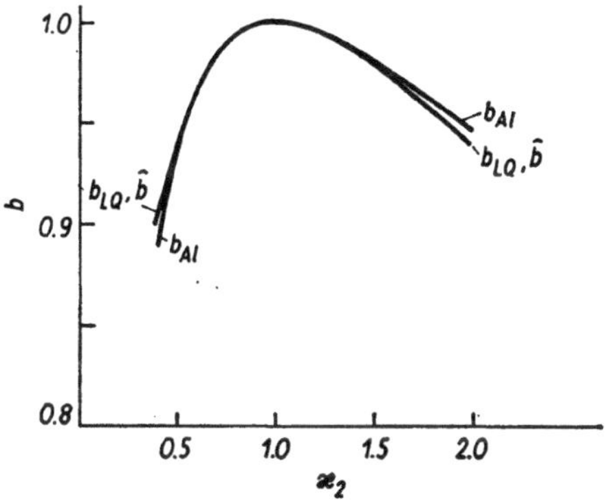

Fig. 33. Comparison of the transmission coefficients b_{AL}, b_{LQ}, and $\hat{b}$

from these curves. Because the equations (4.203), (4.205) and (4.257), (4.258) hold, we now have another explanation as to why the exact solution is approximated fairly well by the approach with F_2. This has already been pointed out by MALISCHEWSKY (1980).

4.4.9.4. Model calculations and comparison of the LQBR-approximation with the exact solution

4.4.9.4.1. General remarks

Using the notation of section 4.8. let us begin with the formulas for the reflection and transmission coefficients of surface waves, which are the basis for the following model calculations. In this context the transmission coefficients refer to the displacement ratio at the surface. For the impedance jump problem, the exact formulas according to Wiener-Hopf are given by

$$|R| = \frac{|k_1 - k_2|}{k_1} \cdot \frac{1}{1 + \varkappa_1}, \qquad |T_r| = \frac{2\sqrt{k_1 k_2}}{k_2} \cdot \frac{\varkappa_1}{1 + \varkappa_1}, \tag{4.314}$$

while the LQBR-approximation

$$\begin{aligned} a_{LQ} &= \frac{k_1^2 - \varkappa_2^2 k_2^2}{k_1^2 + \varkappa_2^2 k_2^2 + k_1 k_2 \left(\varkappa_1 + \dfrac{\varkappa_2^2}{\varkappa_1}\right)}, \\ B_{LQ} &= \frac{k_1(k_1 + \varkappa_2 k_2)}{\varkappa_2 \left\{k_1 k_2 \left(1 + \dfrac{\varkappa_2}{\varkappa_1}\right) + \dfrac{(\varkappa_2 k_2 - k_1)^2}{\varkappa_1 + \varkappa_2}\right\}}, \end{aligned} \tag{4.315}$$

and the "integral-free" terms of the exact solution according to (4.257) and (4.258)

$$\hat{a} = \frac{k_1 - \varkappa_2 k_2}{k_1 + \varkappa_2 k_2}, \qquad \hat{B} = \frac{4k_1\varkappa_1}{(\varkappa_1 + \varkappa_2)(k_1 + \varkappa_2 k_2)} \tag{4.316}$$

refer to a general vertical discontinuity. Further on, we need the phase velocity c of surface waves in a half-space with impedance surface,

$$c = \frac{\omega}{k} = \left(\frac{\varrho}{\mu} + \frac{\varepsilon^2}{\mu^2\omega^2}\right)^{-1/2}, \tag{4.317}$$

which follows from equation (3.37) in a simple way.

In the following, for the sake of brevity we omit the dimensions in most cases. To exclude misunderstandings, let us repeat here once more the dimensions of the most important quantities: $[T] = \mathrm{s}$, $[\omega] = \mathrm{s}^{-1}$, $[c] = \mathrm{km} \cdot \mathrm{s}^{-1}$, $[\varrho] = \mathrm{g} \cdot \mathrm{cm}^{-3}$, $[\mu] = 10^9\ \mathrm{N} \cdot \mathrm{m}^{-2}$, $[\varepsilon] = 10^6\ \mathrm{N} \cdot \mathrm{m}^{-3}$.

4.4.9.4.2. Comparison of approximative and exact solutions of various parameter values

First, let us consider a model with the parameters $\varepsilon = 8.165$, $\mu = 71.54$, $\varrho = 3.3$ and $T = 2\pi$, and we calculate the reflection and transmission coefficients according to (4.314) to (4.316) as a function of $\varkappa_1$ for $\varkappa_2 = \varkappa_3 = 1$. The absolute values of the reflection coefficients are plotted in Fig. 34 and of the transmission coefficients in Fig. 35.

The curves for $|T_r|$, B_{LQ}, and $\hat{B}$ hardly differ from each other. Fig. 36 shows the corresponding phase velocities according to (4.317) right and left of the impedance jump.

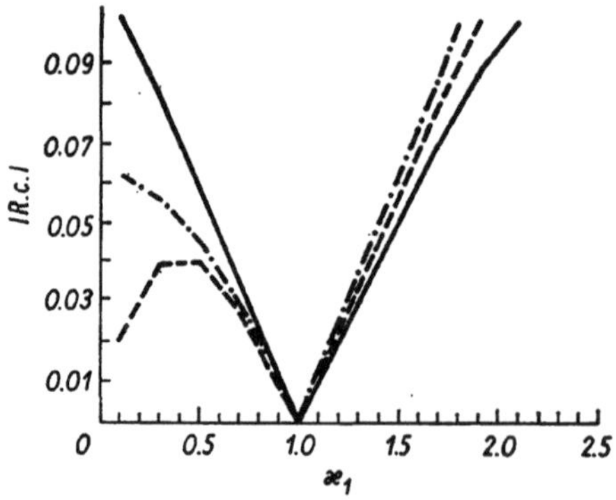

Fig. 34. Comparison of the reflection coefficients $|R|$, $|a_{LQ}|$, and $|\hat{a}|$; $|R|$: full, $|a_{LQ}|$: dashed, $|\hat{a}|$: dash-dotted

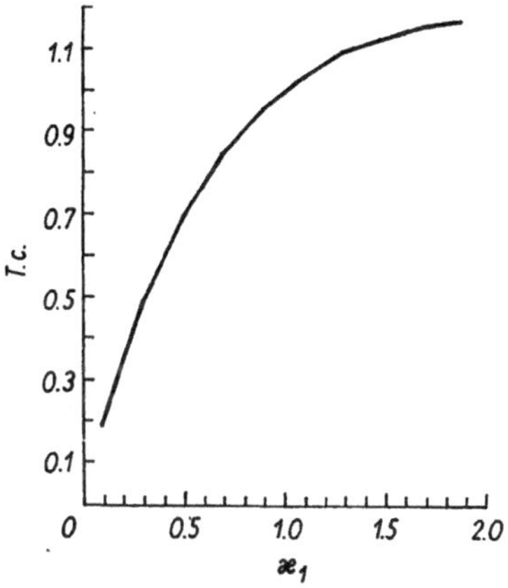

Fig. 35. Comparison of the transmission coefficients $|T_r|$, B_{LQ}, and $\hat{B}$

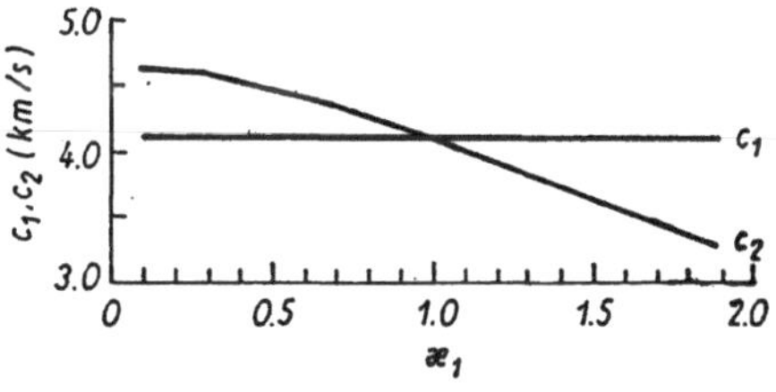

Fig. 36. The phase velocities left (c_1) and right (c_2) of the discontinuity

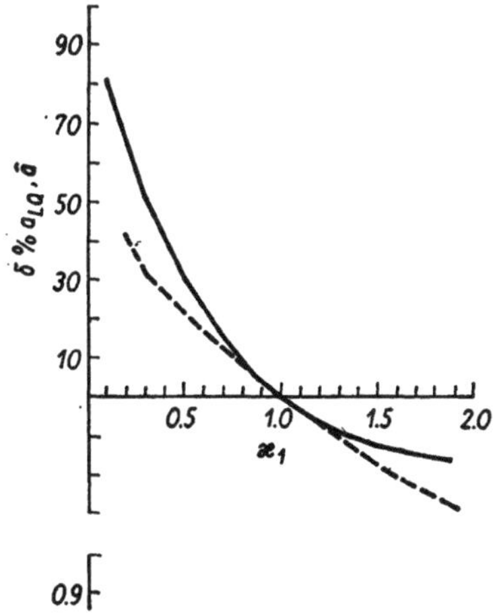

Fig. 37. The percentage errors of a_{LQ} (full) and $\hat{a}$ (dashed) concerning $|R|$

Table 5. The percentage errors of a_{LQ}, B_{LQ}, $\hat{a}$, and $\hat{B}$ for various parameter combinations as a function of $\varkappa_1$; $T = 2\pi$, $\varrho = 1$

ε	μ	μ/ε	$\varkappa_1$	$\Delta a_{LQ}\%$	$\Delta B_{LQ}\%$	$\Delta\hat{a}\%$	$\Delta\hat{B}\%$	c_1/c_2	PEUF
10	1	0.1	0.1	48.6	−6.1	3.6	34.2	0.14	5.9
			0.3	23.8	7.4	1.1	14.7	0.31	1.2
			0.5	10.5	4.4	0.5	5.5	0.51	0.3
			0.7	3.2	1.4	0.2	1.5	0.70	0.07
			0.9	0.3	0.1	0.06	0.14	0.90	0.006
			1.1	0.2	0.1	−0.05	0.11	1.10	0.004
			1.3	1.6	0.8	−0.11	0.84	1.30	0.03
			1.5	3.7	1.8	−0.17	1.99	1.50	0.07
			1.7	6.1	3.0	−0.2	3.38	1.69	0.11
			1.9	8.6	4.0	−0.24	4.88	1.89	0.15
1	1	1	0.1	78.3	0.48	35.7	1.44	0.71	60.0
			0.3	46.5	0.49	25.2	1.14	0.74	24.9
			0.5	25.4	0.54	16.2	0.69	0.79	9.2
			0.7	11.6	0.25	8.8	0.27	0.86	2.5
			0.9	2.9	0.03	2.6	0.03	0.95	0.2
			1.1	−2.2	0.03	−2.4	0.03	1.05	0.2
			1.3	−4.7	0.26	−6.5	0.27	1.16	1.2
			1.5	−5.6	0.67	−9.9	0.73	1.27	2.8
			1.7	−5.4	1.19	−12.8	1.37	1.39	4.6
			1.9	−4.5	1.74	−15.2	2.14	1.52	6.4
0.1	1	10	0.1	81.8	−0.0001	44.9	0.0003	0.995	66.9
			0.3	53.7	0.0001	34.9	0.0003	0.995	29.0
			0.5	33.2	0.0001	24.9	0.0002	0.996	11.1
			0.7	17.5	0.00008	14.9	0.00008	0.997	3.1
			0.9	5.2	0.00001	5.0	0.00001	0.999	0.3
			1.1	−4.7	0.00001	−5.0	0.00001	1.001	0.2
			1.3	−12.9	0.0001	−14.8	0.0001	1.003	1.7
			1.5	−19.6	0.0004	−24.6	0.0005	1.006	4.0
			1.7	−25.3	0.0009	−34.4	0.001	1.009	6.7
			1.9	−30.2	0.002	−44.1	0.002	1.013	9.6

The most remarkable fact is the different size of deviation between the approximations and exact solutions for the reflection and transmission coefficients, respectively. Furthermore, the asymmetry for $\varkappa_1 < 1$ and $\varkappa_1 > 1$ is noticeable, which is also reflected in the percentage errors of a_{LQ} and $\hat{a}$ concerning $|R|$ (Fig. 37) and of B_{LQ} and $\hat{B}$ concerning $|T_r|$ (Fig. 38) respectively.

The phase angles φ_R and φ_T according to (4.247) and (4.250) are plotted in Fig. 39 as a function of $\varkappa_1$.

Table 5 (cont.)

ε	μ	μ/ε	$\varkappa_1$	$\Delta a_{LQ}\%$	$\Delta B_{LQ}\%$	$\Delta\hat{a}\%$	$\Delta\hat{B}\%$	c_1/c_2	PEUF
100	10	0.1	0.1	41.3	−4.6	0.44	41.4	0.1	0.7
			0.3	22.6	7.9	0.12	15.6	0.3	0.13
			0.5	10.1	4.5	0.05	5.7	0.5	0.03
			0.7	3.0	1.5	0.02	1.6	0.7	0.008
			0.9	0.3	0.1	0.01	0.1	0.9	0.0006
			1.1	0.2	0.1	−0.01	0.1	1.1	0.0004
			1.3	1.7	0.8	−0.01	0.9	1.3	0.003
			1.5	3.8	1.9	−0.02	2.0	1.5	0.007
			1.7	6.1	3.0	−0.02	3.4	1.7	0.01
			1.9	8.8	4.0	−0.02	4.9	1.9	0.02
10	10	1	0.1	66.3	−4.3	16.4	14.6	0.32	28.3
			0.3	31.4	4.2	8.2	8.9	0.42	8.8
			0.5	14.0	3.1	4.1	4.0	0.56	2.7
			0.7	4.9	1.1	1.9	1.2	0.73	0.6
			0.9	0.8	0.1	0.5	0.1	0.91	0.5
			1.1	−0.2	0.1	−0.4	0.1	1.09	0.04
			1.3	0.6	0.7	−1.1	0.7	1.28	0.3
			1.5	2.4	1.6	−1.6	1.8	1.46	0.6
			1.7	4.5	2.7	−1.9	3.1	1.65	1.0
			1.9	6.8	3.6	−2.2	4.5	1.84	1.3
1	10	10	0.1	81.4	−0.009	43.7	0.03	0.95	66.8
			0.3	52.8	0.01	33.6	0.02	0.96	28.9
			0.5	32.2	0.01	23.7	0.02	0.97	11.1
			0.7	16.7	0.007	14.0	0.007	0.98	3.1
			0.9	4.9	0.001	4.6	0.001	0.99	0.3
			1.1	−4.3	0.001	−4.5	0.001	1.01	0.2
			1.3	−11.3	0.01	−13.3	0.01	1.03	1.7
			1.5	−16.8	0.03	−21.6	0.04	1.06	4.0
			1.7	−21.0	0.07	−29.7	0.09	1.08	6.6
			1.9	−24.1	0.1	−37.3	0.14	1.11	9.5

Obviously, φ_T can be neglected compared with φ_R expressing the fact that T_r is an essentially real (and positive) quantity, so that surface waves do not undergo phase jumps if they are travelling from one medium into another. In Table 5, the error trends concerning the exact values of the quantities a_{LQ}, $\hat{a}$, B_{LQ}, and $\hat{B}$ are given for different parameter combinations ε and μ. Furthermore we have added the quantity PEUF formed with the exact values $|R|$ and $|T_r|$ as well as the ratio of the phase velocities c_1/c_2.

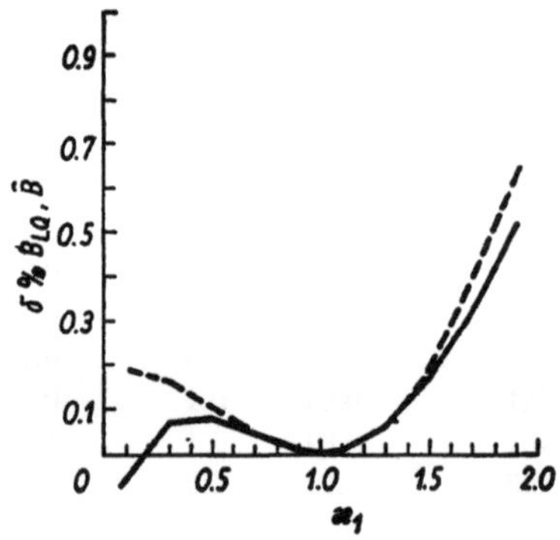

Fig. 38. The percentage errors of B_{LQ} (full) and $\hat{B}$ (dashed) concerning $|T_r|$

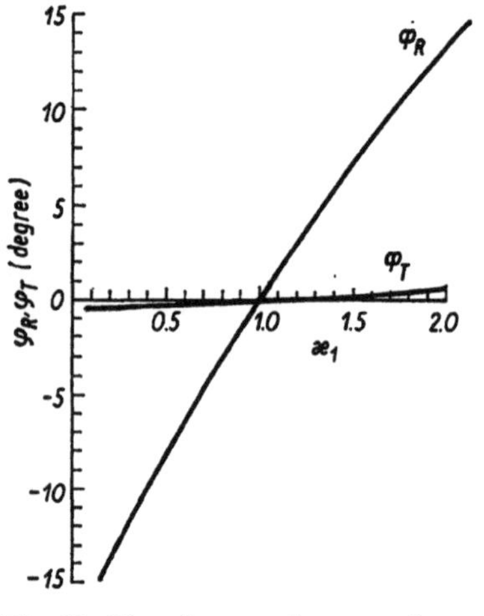

Fig. 39. The phase angles φ_R and φ_T as a function of $\varkappa_1$

The table contains various trends, but here we can only speak about the most important ones. A very characteristic feature is expressed by the fact that in practically the whole range, the errors of the approximated transmission coefficients are clearly smaller (partly even by orders of magnitude) than those of the reflection coefficients. Furthermore, the exact transmission coefficient T_r is better approximated by B_{LQ} than by $\hat{B}$. These two trends are more or less independent of PEUF. However, the error of the reflection coefficients depends considerably on the magnitude of PEUF, i.e. with increasing PEUF, the error increases as well. This will also be true if PEUF is formed from the approximated coefficients in the general case when the exact ones are not available. For constant μ and with decreasing ε the error of a_{LQ} increases while that of B_{LQ} decreases. Since the absolute value of the reflection coefficients decreases with decreasing ε, we can state that the weaker these ones are approximated by the LQBR-method the smaller they are for the same impedance contrast (i.e. $\varkappa_1$). On the other hand, the transmission coefficients belonging to the small reflection coefficients are approximated especially well. The ratio μ/ε seems to be a characteristic quantity, and in the following we try to explain this fact heuristically. Roughly speaking, the different behaviour of the amplitude-depth functions $f_1 \sim \exp\left(-\frac{\varepsilon_1}{\mu} x_3\right)$ and $f_2 \sim \exp\left(-\frac{\varepsilon_2}{\mu} x_3\right)$

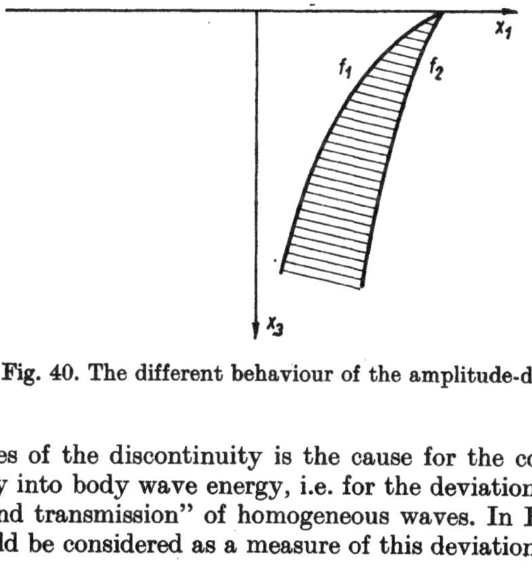

Fig. 40. The different behaviour of the amplitude-depth functions

at both sides of the discontinuity is the cause for the conversion of surface wave energy into body wave energy, i.e. for the deviation from the "classical reflection and transmission" of homogeneous waves. In Fig. 40, the hatched area F_ε could be considered as a measure of this deviation and in our case it is given by

$$F_\varepsilon = \frac{\mu}{\varepsilon} \cdot \frac{1 - \varkappa_1}{\varkappa_1} \qquad \text{(compare with 4.134!).} \qquad (4.318)$$

Obviously, F_ε increases with decreasing ε and with increasing μ, and that explains principally the error trend of a_{LQ}. On the other hand, the opposite behaviour of B_{LQ} cannot be explained in this way. The asymmetrical function $\frac{1}{\varkappa_1} - 1$ in (4.318) explains up to a certain degree the asymmetrical behaviour for $\varkappa_1 < 1$ and $\varkappa_1 > 1$, respectively.

To summarize we can state that the transmission coefficients are approximated very well by the least squares method, but for the reflection coefficients, being the more sensitive quantities by far, this is only true for sufficient small impedance jumps. For the latter ones, the reflection coefficient has the simple form [compare with RULF (1977a)]

$$R = \frac{c_2 - c_1}{c_2 + c_1}, \tag{4.319}$$

which is valid, incidentally, for arbitrary impedance jumps in the case $\varepsilon \to \infty$. Furthermore, it has to be noted that for small PEUF, considerable errors of the reflection coefficient may also occur.

Finally note that we have to expect modifications, if there are several modes, i.e. if the half-space is inhomogeneous in x_3-direction. However, in the author's opinion, the essential fundamental ideas may also be applied to this case.

4.4.9.4.3. Simulation of the period dependence

First, let us discuss that period dependence which is intrinsic for the model under consideration. According to (4.317) the surface waves of the impedance surface (in contrast to the seismic surface waves) show anomalous dispersion and for $T \to 0$ the shear wave velocity $\beta = \sqrt{\mu/\varrho}$ is adopted. For the approximated and exact reflection and transmission coefficients the period dependences are given in Fig. 41 and Fig. 42 which refer to an impedance jump with $\varkappa_1 = 1.5$ and to the same model parameters as in Fig. 34. The curves for B_{LQ} and $\hat{B}$ hardly differ from $|T_r|$. The corresponding dispersion curves $c_1(T)$ and $c_2(T)$ are plotted in Fig. 43.

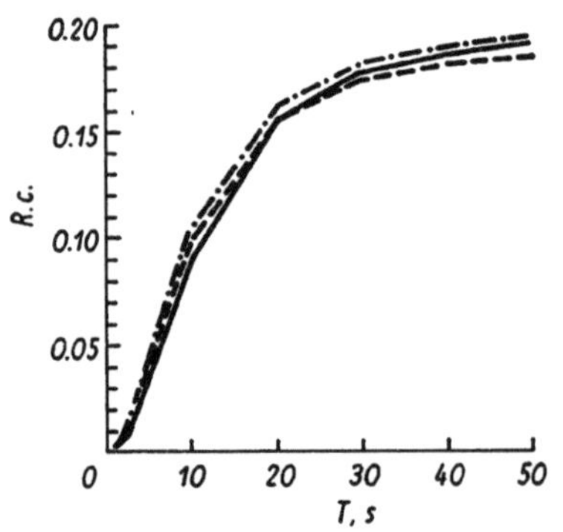

Fig. 41. T-dependence of reflection coefficients; $|R|$: full, a_{LQ}: dashed, $\hat{a}$: dash-dotted

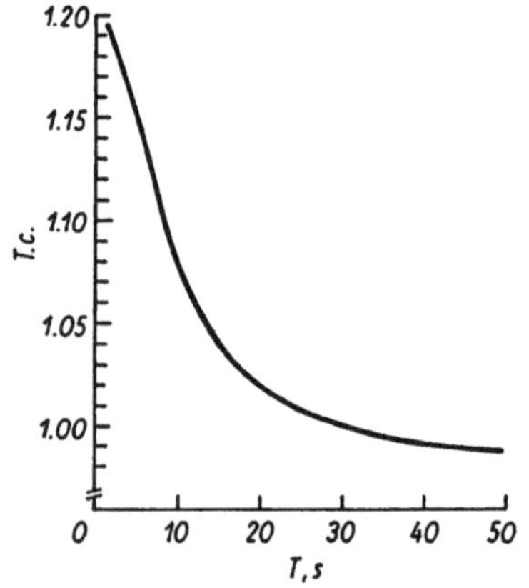

Fig. 42. T-dependence of transmission coefficients

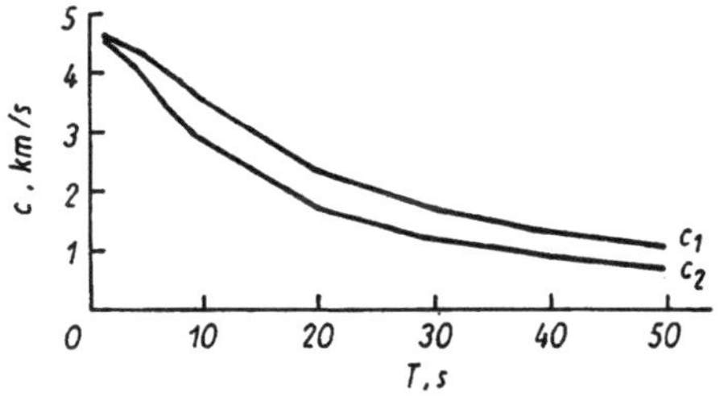

Fig. 43. The phase velocity dispersion curves $c_1(T)$ and $c_2(T)$

Since the amplitude-depth functions do not depend on period here, T is not a good parameter to simulate the corresponding phenomena for seismic surface waves, where the depth of penetration is controlled by T. However, this T-dependence (that is depth of penetration) can be simulated by varying the "impedance level" ε_1 for constant $\Delta\varepsilon$ (that is material contrast). Define the new parameters Θ_1 and Θ_2 as reciprocal values of ε_1 and ε_2 respectively, and it follows that

$$\frac{1}{\Theta_2} = \frac{1}{\Theta_1} + \Delta\varepsilon \tag{4.320}$$

and

$$\varkappa_1 = 1 + \Theta_1 \Delta\varepsilon. \tag{4.321}$$

Consequently, the eigenfunctions left and right of the discontinuity are proportional to $\exp(-\mu/\Theta_1)$ and $\exp\left[-\mu\left(\frac{1}{\Theta_1} + \Delta\varepsilon\right)\right]$, respectively, so that the

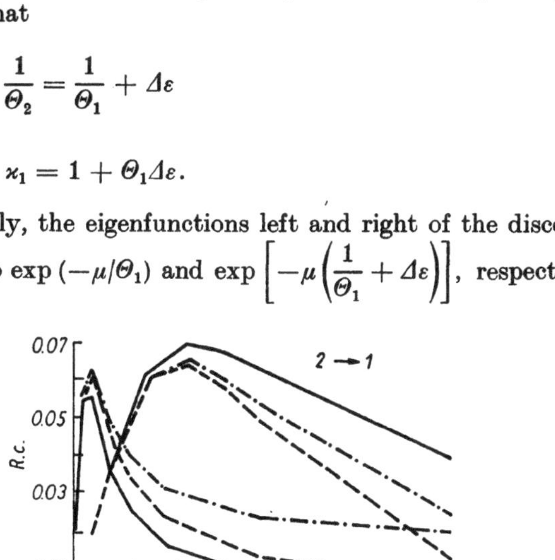

Fig. 44. Simulation of the period dependence of reflection coefficients; $|R|$: full, a_{LQ}: dashed, $\hat{a}$: dash-dotted
$1 \to 2$: $\Delta\varepsilon = 4$, $2 \to 1$: $\Delta\varepsilon = -4$; μ and ϱ as in Fig. 34

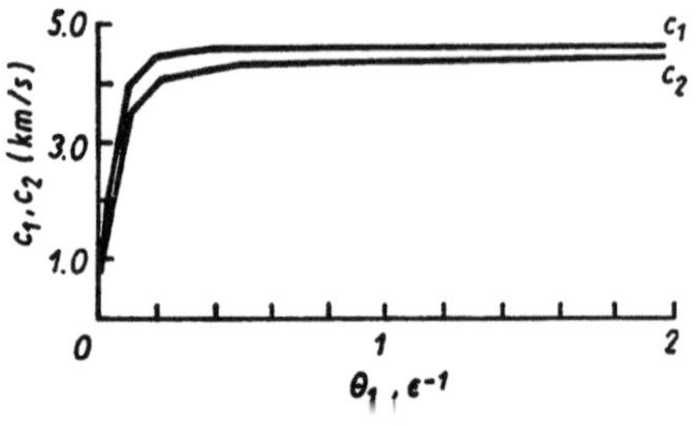

Fig. 45. Simulated dispersion curves for the direction of propagation $1 \to 2$

depth of penetration increases with increasing Θ_1 analogously to the T-behaviour of seismic surface waves. Using the parameter Θ_1 we simulate now the period dependence of the reflection coefficient analogously to seismic surface waves. For both directions of propagation, the exact and approximated reflection coefficients are represented in Fig. 44 as a function of Θ_1. The corresponding dispersion curves for direction $1 \to 2$ are shown in Fig. 45.

Note in Fig. 44 that, regardless of the, in some cases, considerable deviations of the approximated coefficients from the exact ones, the occurring characteristic maximum is reproduced fairly well. This is an important fact with regard to seismology.

Using the theory developed here we simulate now the propagation of Love waves across a continental margin. Roughly simplifying, the continental margin could be thought of as consisting of models 1 and 2, whose parameters are given in Table 6. The quantities $H_{(1)}$ and $H_{(2)}$ refer to the layer thicknesses of the first and second layer respectively.

For the Love wave fundamental mode, the corresponding phase velocity dispersion curves are shown in Fig. 46. Here, the period of Love waves is denoted by T', while the period of surface waves of the impedance surface continues being $T = 2\pi$. The T'-dependence of the reflection and trans-

Table 6. Simplified model of a continental margin

	Model 1	Model 2
β_1 (km/s)	3.723	3.723
ϱ_1 (g/cm^3)	2.77	2.77
H_1 (km)	30.	10.
β_2 (km/s)	4.656	4.656
ϱ_2 (g/cm^3)	3.3	3.3
H_2 (km)	∞	∞

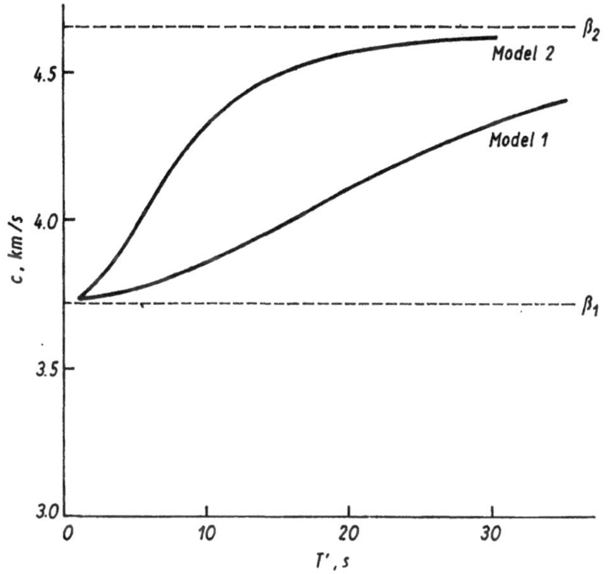

Fig. 46. The phase velocity dispersion curves for the media left and right from the continental margin

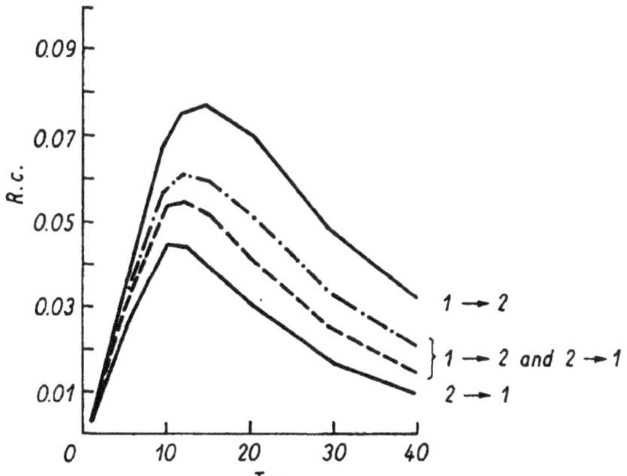

Fig. 47. Simulated period dependence of the reflection coefficients for the continental margin for both directions of propagation; $|R|$: full, a_{LQ}: dashed, $\hat{a}$: dash-dotted

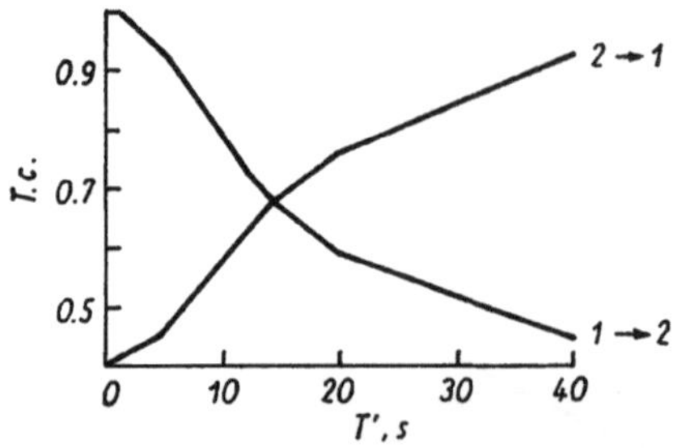

Fig. 48. Simulated period dependence of the transmission coefficients for the continental margin for both directions of propagation

mission coefficients we are interested in, can be investigated approximately by varying ε (and $\varkappa_1$) for different T' as long as the corresponding phase velocities from Fig. 46 are obtained, and then the coefficients are determined. Both directions of propagation (model 1 $\rightarrow$ model 2 and vice versa) were considered and the corresponding reflection and transmission coefficients are plotted in Fig. 47 and Fig. 48, respectively, as a function of T'.

For this configuration, the occurrence of a characteristic maximum for the reflection coefficient is theoretically plausible. Furthermore, the characteristic, different trend of the transmission coefficients for the two directions of propagation, agrees with the observed data as well as with other theoretical calculations [see e.g. GREGERSEN and ALSOP (1974)]. Again, the maximum of the reflection coefficient is shifted only slightly for the approximations, and the approximated and exact transmission coefficients hardly differ.

4.4.9.4.4. Simulation of a vertical discontinuity (contrast in μ and ϱ)

In this section, we simulate a proper vertical discontinuity by the impedance jump model using the relations (4.269) and (4.271). The two equivalent configurations are shown in Fig. 49.

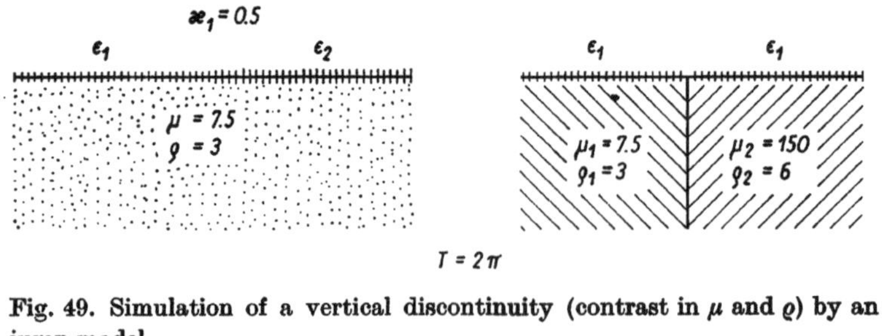

Fig. 49. Simulation of a vertical discontinuity (contrast in μ and ϱ) by an impedance jump model

The equivalence refers to the reflection and transmission coefficients and the phase velocities of the surface waves as well. The reflection and transmission coefficients are plotted as functions of ε in Fig. 50 and Fig. 51, respectively.

Note that the errors of the approximated transmission coefficients are again considerably below those of the approximated reflection coefficients. Of course, we are also interested in the dependence of the coefficients on the material contrast itself, i.e. on $\varkappa_2$ for instance. That is why we have carried out corresponding calculations, and the results are shown in Fig. 52. Certainly, we cannot present the exact solution, because it is no longer known in this case.

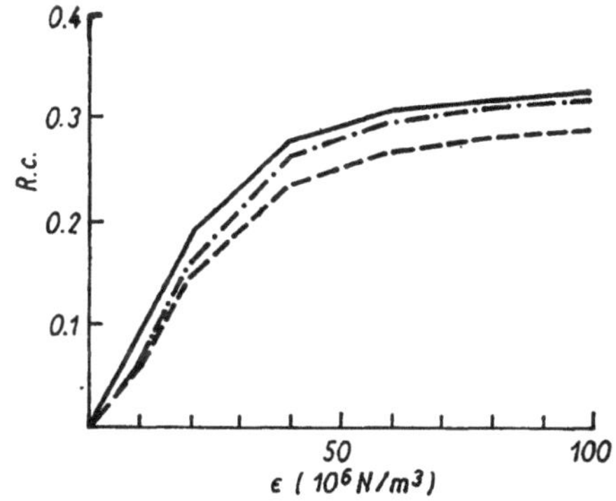

Fig. 50. The reflection coefficients as a function of ε
$|R|$: full, a_{LQ}: dashed, $\hat{a}$: dash-dotted

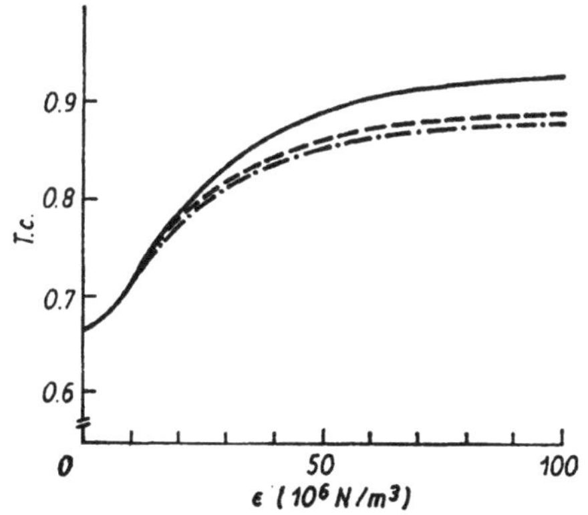

Fig. 51. The transmission coefficients as a function of ε; $|T_r|$: full, b_{LQ}: dashed, $\hat{b}$: dash-dotted

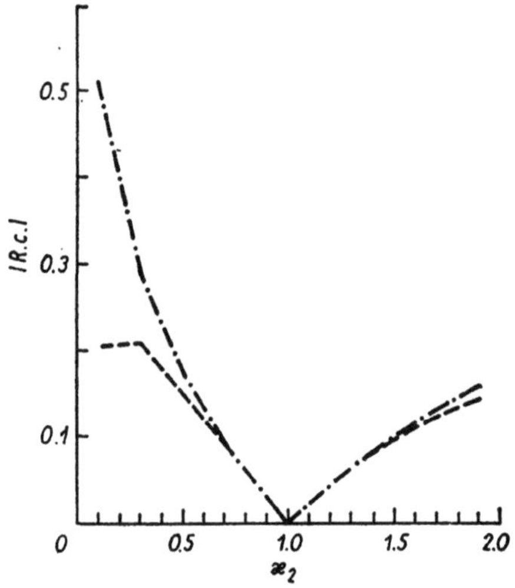

Fig. 52. The reflection coefficients as a function of $\varkappa_2$
a_{LQ}: dashed, $\hat{a}$: dash-dotted
model parameters: $T = 2\pi$, $\varepsilon = 1$, $\mu = 75$, $\varrho = 3$

4.4.10. Discussion of further attempts at solution in literature

We cannot discuss all attempts at solution of the problem here. We will only mention some further papers about wave-guides with discontinuities from the non-seismological literature which are important in the author's opinion. Let us start with some papers dealing with elastic waves. It is widely known that thin strips of one material deposited on a substrat of another material or slits in the substrat are capable of guiding surface waves. Numerous surface acoustic wave devices take advantage of this fact as already mentioned. FREUND (1971) considered guided surface waves on an elastic half-space and the reflection of these at an obstacle as a Wiener-Hopf problem using the Laplace transformation. He refers to the fact that the limited range of applicability of the Wiener-Hopf method also limits the number of surface wave reflection problems which can be solved in detail. The Wiener-Hopf method was also applied by YONEYAMA and NISHIDA (1974) to obtain a first order approximation for the reflection and transmission of Rayleigh waves at the edge of a thin film deposited on the substrat. It was assumed that the film is thin enough in comparison with the wave length so that Tiersten's boundary conditions (see section 2.1.2.) could be applied. SIMONS (1975, 1976) solved the same problem for both Love and Rayleigh waves using the Fourier transformation and the Wiener-Hopf method, but the boundary conditions at the edge of the film were formulated more rigorously.

As you know, any change of the character of the surface, along which the surface waves propagate, leads to reflections and causes a conversion of the energy into other wave types. In this respect, the reflection of surface waves at the edge of a thin film is a crucial problem. According to MATTHEWS (1977) there are on the whole two quantities which have an essential influence on the reflection, firstly the wave length λ of the surface waves and secondly the height of the step or the film thickness h, so that the reflection coefficient r, depending on the ratio of these two quantities, may be expanded in a Taylor series as follows

$$r(h/\lambda) = r_0 + C(h/\lambda) + O(h/\lambda). \tag{4.322}$$

For certain configurations, r_0 vanishes identically, so that r is then, on the whole, a linear function of h/λ.

The reflection of Rayleigh waves from groove and step discontinuities was investigated by TUAN and LI (1974) using a perturbation method. Li developed an approximation method to solve similar problems on the basis of equivalent electrical networks. Using this approach, LI and MELNGAILIS (1975) succeeded in investigating the interesting problem of energy storage at step discontinuities. The concept of energy storage at discontinuities is common in microwave physics but, astonishingly, this is hardly the case in the theory of elasticity. It is widely known that from a certain limiting period, SH-modes can no longer propagate in a plate, i.e., if they are generated at a discontinuity, they cause an "energy storage" in its vicinity. However, it is less known, that the same effect may also occur for example at a step discontinuity on the surface of an elastic half-space. In this case, the energy storage refers to the continuous spectrum. The body waves, generated and radiated from a discontinuity, belong to that part of the continuous spectrum which is analogous to the waves radiated away from an antenna into the "visible" region. The remainder of the continuous spectrum, however, is analogous to the waves of the "invisible" region representing the storaged reactive energy in the vicinity of an antenna. From experience it is known that energy storage causes phase changes of the reflected and transmitted waves at a discontinuity. These phase changes are important for the outline of surface wave devices, but they have to be taken into account for the interpretation of seismic surface waves as well.

The conversion of surface waves into body waves which takes place at a discontinuity, is important for the tapping and feeding of surface wave energy from or into a wave-guide. It was investigated e.g. by CHANG and TUAN (1972, 1973) for Rayleigh and Love waves as well. The discontinuity was considered as a local deviation of the thickness of the film deposited on the substrat, and a perturbation scheme together with Green's function technique was applied. Because there are no infinite half-spaces, but plates of finite thickness at real surface wave devices, the body waves generated at a discontinuity couple with the corresponding plate modes. This coupling was investigated experimentally as well as theoretically for example by HAUS et al. (1980). The discontinuity problem for the elastic wave-guide was also considered by KOSHIBA et al. (1981) using a combined half-analytical method. In this paper, a finite-element method is used for the interior region enclosing the discontinuities, while an analytical method is utilized for the exterior region (the uniform wave-guides). In this connection, let us finally quote the recent work of SHEN et al. (1984). Here, the transmission and reflection of surface acoustic waves by a transverse boundary separating two isotropic media (with finite thicknesses) had been studied by the collocation method and the least-squares boundary residual method. The authors are able to illustrate the influence of the material properties and plate thicknesses on the surface reflection and transmission coefficients, as well as the efficiencies and radiation patterns for the excited bulk waves. All of the coefficients depend on both the sign and magnitude of the acoustic impedance mismatch across the interface (this should not be confused with our "impedance" parameter ε!). Further on, the conversion into the reflected bulk waves is far more efficient than into the transmitted bulk waves for all cases studied. Although the reflected and transmitted surface waves are relatively insensitive to Poisson's ratio, the radiation patterns of the reflected and transmitted bulk waves vary substantially with the choice of this parameter. Finally, it was found that the mass density dependence of the acoustic impedances may be the most important parameter for the magnitude but not the angular dependence of the radiation patterns.

The scattering of Rayleigh waves at a surface-breaking crack was solved exactly by MENDELSOHN et al. (1980) by applying an integral transformation technique and transforming the boundary value problem into two uncoupled singular integral equations, and the latter were solved numerically using the collocation method. This exact solution does, however, involve a substantial computational effort and that is why ACHENBACH et al. (1980) have proposed an approximate solution based on elastodynamic ray theory. It is valid in the high-frequency range and is also suitable for the solution of the inverse problem. We cannot cite here numerous further papers about the scattering of surface waves at various surface defects. But we want to refer to the concept of so-called surfons introduced by Ezawa [see e.g. EZAWA et al. (1974)] for the solution of similar problems. Surfons are quanta of elastic waves (analogous to the phonons), which satisfy the boundary conditions on the surface of a solid body. They form a complete system and, consequently, they cover surface and body waves as well.

Now we will discuss some recent papers about discontinuities in electromagnetic wave-guides. STEGEMAN et al. (1981) investigated the refraction of surface polaritons at a vertical interface in a closed wave-guide by satisfying the boundary conditions at 52 equally spaced points and using 52 modes from each side of the boundary. The procedure of matching the boundary conditions always ensures that they are satisfied at each of the discrete set of points. It does not guarantee, however, that they are also satisfied elsewhere. The criterion used to check the validity of the solutions was energy conservation. In most cases, the latter was excellent, but it was very sensitive to the completeness of the set of the first 52 modes used for the given configuration. If a mode was missing, especially from the first 15—20 modes, the energy conservation worsened considerably. Perhaps the most surprising aspect of the numerical results in this paper is the large fraction of the incident energy that is transmitted through the boundary. In other words, very large transmission and very small reflection coefficients were obtained in comparison with the reflection and refraction of homogeneous, plane waves. The investigated model is shown in Fig. 53.

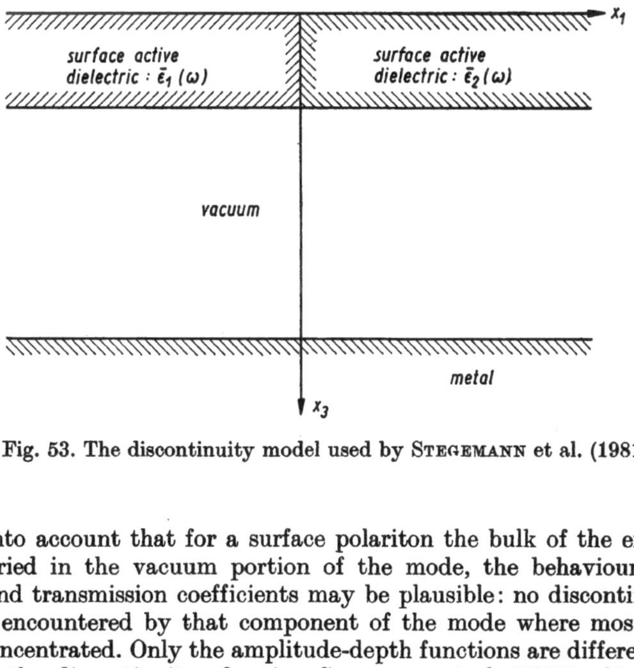

Fig. 53. The discontinuity model used by STEGEMANN et al. (1981)

Taking into account that for a surface polariton the bulk of the energy is usually carried in the vacuum portion of the mode, the behaviour of the reflection and transmission coefficients may be plausible: no discontinuity in material is encountered by that component of the mode where most of the energy is concentrated. Only the amplitude-depth functions are different right and left of the discontinuity. Quoting STEGEMAN et al. (1981), this lack of discontinuity in material properties in the region where most of the mode energy is present may be the deciding factor that produces the large transmission and small reflection coefficients.

In a former paper, the scattering of surface waves at an impedance jump considered by us was also solved exactly by KAY (1959) using a modified version of the Wiener-Hopf procedure. This exact solution was applied by YONEYAMA and NISHIDA (1981) using the following argumentation. The latter authors searched for a possibility which was as practicable as possible to determine the reflection and transmission coefficients of surface waves for a wave-guide with a step discontinuity (see Fig. 54). Certainly, the method which is based on the discretization of the radiation modes by polynomials yield a precise solution, but it takes much computation time. That is why YONEYAMA and NISHIDA (1981) propose to apply KAY's (1959) exact results which are available without computer aid to a dielectric slab wave-guide by analogy. To do this, KAY's (1959) result for impedance-like boundary conditions have to be transformed into the corresponding terms of the dielectric wave-guide. The results obtained in this way were compared with ROZZI's (1978) exact solution, and the agreement was quite good, especially when the difference of the dielectric constants across the step discontinuity was small.

The scattering of surface waves at a step discontinuity of a dielectric slab wave-guide (see Fig. 54) was treated by ITTIPIBOON and HAMID (1979) using an alternative method.

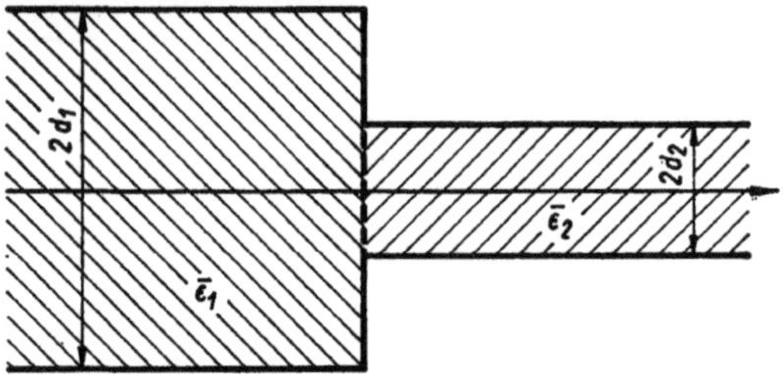

Fig. 54. The step discontinuity in the dielectric slab wave-guide

Applying the mode-matching procedure according to MARCUSE (1970) two simultaneous equations are obtained. However, instead of using a perturbation technique to solve these equations, the residue-calculus technique is applied [see e.g. MITTRA and LEE (1971)]. Essentially, a complex function is constructed so that its poles have the same locations and contributions as those of the surface wave poles, whilst its branch cut integral is equal to the contribution from the radiated field. ITTIPIBOON and HAMID (1981) also solved the same problem for the special case $d_1 = d_2$ using the Wiener-Hopf technique. However, for this case the factorization cannot be carried out in a closed form. On the contrary, the solution contains unknowns which satisfy an infinite, linear, algebraic system of equations. But it was shown that a satisfactory agreement with the residue-calculus results can be obtained by using only a few terms of this system, e.g. $N = 3$. The essential numerical statement of this paper is, that the reflection coefficient increases with frequency up to a certain value and then decreases to the reflection coefficient of the normal plane wave incidence at the junction of two media of infinite dimensions. This behaviour is physically plausible as a result of the skine effect. It was also found that the reflection coefficient is increased when $\bar{\varepsilon}_2$ is increased with respect to $\bar{\varepsilon}_1$. The normalized transmitted power was found to increase with frequency as well as with the ratio $\bar{\varepsilon}_2/\bar{\varepsilon}_1$. When the thickness on both sides of the junction is not the same, the formulation of the problem by the Wiener-Hopf technique is too complicated and it is more convenient to apply the residue-calculus technique for example. If the factorization could somehow be carried out analytically, the complexity and the computer time would be essentially reduced.

By applying a spectral expansion MANENKOV (1978) approximately solved the problem concerning the propagation of surface waves along an inhomogeneous part of a thin dielectric wave-guide. GONCHARENKO and KARPENKO (1983) proposed an approximation method for the treatment of the scattering of surface waves at the free end of a wave-guide. It is based on the assumption that for the reflection of a normal mode, the bulk of the energy remains in the wave-guide and only a small amount is scattered backwards into radiation modes. Furthermore, it is noted that the transformation of an incident normal mode into waves of the discrete and continuous spectra is in physical terms one and the same process. That is why it is natural to assume that the energies of the produced radiation modes and of the normal modes with numbers $\nu \neq \mu$ (let ν be the number of the incident mode) are of the same order of magnitude.

Using a completely different method, the same problem, namely the scattering of surface waves in an abruptly ended slab dielectric wave-guide, was also solved by GELIN et al. (1981). It consists of deriving a system of two coupled singular integral equations of discrete and continuous wave amplitudes interfering in the total electromagnetic fields which are matched at the discontinuity interface. Note that we previously obtained one singular integral equation only [see section 4.4.3.]. A straightforward approach for solving the coupled integral equation system is achieved by means of the standard Neumann series. A first order solution is derived by neglecting the reflected discrete and continuous modes, and this is substituted into the second integral equation to obtain an improved solution, and so on. The series are found to converge rapidly. The application of this technique to other kinds of discontinuities, e.g. to the step discontinuity, was also successful.

KUDO and MUSHIAKE (1982) treated the reflection and transmission of guided waves in the case where a three-layer wave-guide is joined with a four-layer wave-guide. The analysis is analogous to RULF and KEDEM (1976), but we cannot follow the authors as to how it should be possible to reduce the reflection and transmission coefficients to simple integrations only, without the constraint of solving an integral equation.

4.5. Lateral inhomogeneities and seismic surface waves

4.5.1. *General remarks*

Lateral inhomogeneities between source and receiver may influence the records of seismic surface waves considerably, which is why they have to be taken into account for the interpretation. On the other hand, the interpretation of surface wave records yields geologically and, as a result, economically valuable information about the location as well as about the depth penetration of vertical and sloping discontinuities. There are a very large number of papers devoted to the interpretation of surface wave seismograms with regard to the depth dependence of the elastic parameters. But up to now, only a few papers have dealt with the identification and localization of lateral discontinuities on the basis of seismic surface waves. As a rule, body wave seismograms are characterized by sharp arrivals which have their origin in reflections from various interfaces within the Earth. The surface wave seismograms also have a structure, which has to be "decoded", and where more or less separated wave groups dominate without sharp arrivals, since dispersion is present as a complicating factor. In the following, we mention only those contributions which were or which are essential in this field or where the aim is pronounced especially clearly in the sense mentioned above.

4.5.2. *The interpretation of surface waves in media with lateral inhomogeneities*

Here, as in similar cases, it is possible to make statements without having solved the complicated boundary value problem concerning the passage of surface waves through a discontinuity. First, it all depends on the identification of the reflected (multiple) wave groups in the seismogram. The formation of these wave groups depends to a high degree on the geological conditions near the station and in the source region as well. In certain suitable cases, the identification of the multiple wave groups is already possible using the analog records [see e.g. SIKHARULIDZE (1967), MURPHY and O'BRIEN (1978), and NEUNHÖFER (1985)].

In Fig. 55, we show seismograms of a near seismic event recorded at the stations Berggießhübel (BRG) and Pritzwalk (PRW) in the GDR. Noteworthy is the different character of the seismograms. In PRW we recognize a Rayleigh wave group between the minutes 4 and 5 which was probably caused by reflections from fault zones in the Hartz Mountains region. These reflections are not observed at stations south of the source (e.g. BRG). Consequently they cannot occur north of the source but have to occur south of it (Fig. 56).

The monography of SIKHARULIDZE (1978) contains numerous further exam-

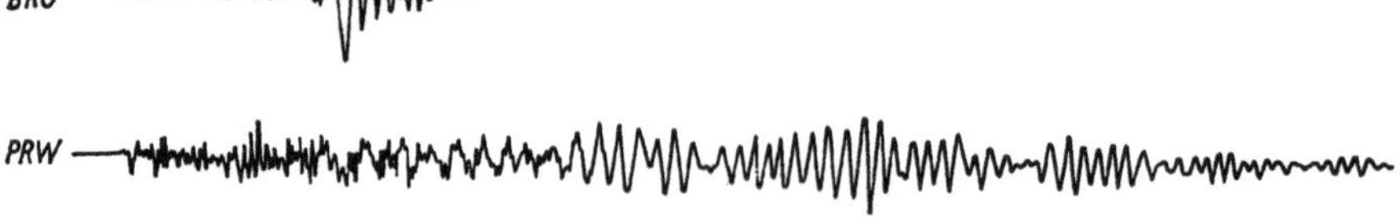

Fig. 55. Seismograms of a near seismic event in the GDR

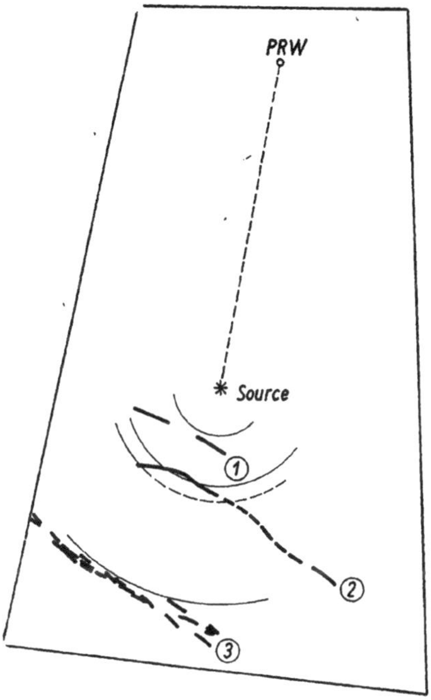

Fig. 56. Reflecting elements for surface waves of a near seismic event

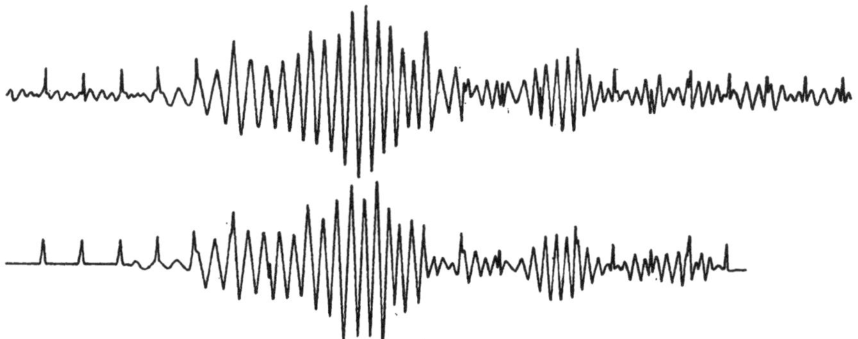

Fig. 57. Splitting of an seismogram into the direct (middle) and reflected (below) Rayleigh fundamental mode, top: original seismogram of an earthquake from the Mid Atlantic ridge recorded at Moxa (MOX)

ples concerning the territory of the USSR. He has pointed out that the correct identification of such wave groups requires great efforts and experience from the interpreter. Generally, a digital data processing is necessary. The following example concerning an earthquake from the Mid Atlantic ridge will illustrate this. As you can see in Fig. 57, using time-dependent filtering it was possible to isolate besides the Rayleigh fundamental mode, a later arriving, i.e. reflected wave group [see also Malischewsky and Neunhöfer (1983)].

Besides the visual separation of reflected and direct waves, Murphy and O'Brien (1978) also applied a refined digital analysis to determine the azimuth of the reflected waves. For this purpose, the seismograms of the little array were subjected to a velocity and frequency filtering. The selected velocity filter was a simple beam steer, where the usual summation and delay processes were applied. The frequency filtering consisted of a bandpass filter which eliminated the shorter period body wave phases.

Sikharulidze (1978) proceeds from the simplifying assumption that the forms of oscillation of the direct and reflected waves are similar one to another, and on the whole they differ by their amplitudes only, where the amplitude changes are caused by reflection, transmission, and different directions of propagation. However, it has to be noted that quite apart from the mode conversions at the interface, the direct wave may be modulated by the reflected one [see Drake (1980)]. This effect will be the more pronounced, the nearer the station is located to the discontinuity, and it has to be taken into account for the mode elimination. Note that the theoretically calculated reflection and transmission coefficients are automatically valid for a well-defined mode. Proceeding from the real seismogram, a mode elimination has to be carried out beforehand to determine the corresponding coefficients. For Love wave propagation across a step discontinuity, Martel (1980) applied a spatial filtering proposed by Cara, which allows the isolation of a normal mode if its phase velocity dispersion curve is known.

To identify reflected wave groups in more complicated cases, Sikharulidze (1978) carried out both an autocorrelation and a spectral analysis of the direct and reflected waves. Generally, the essential step in interpreting the observed data is the development of schemes which give an idea of the structure and the properties of the reflecting and refracting surface. To interpret the data correctly, it is useful, and in some cases necessary, to have additional informations about the discontinuities from other geophysical and geological considerations. Certain investigations can be carried out on the basis of reflected and refracted surface waves alone. Among them there is the determination of the reflecting elements which is possible in suitable cases using the so-called *method of ellipses* [see e.g. Sikharulidze (1978)]. Proceeding from the assumption that the reflected wave group has been successfully separated from the direct wave group, the reflection has to take place on that curve which is the geometrical locus of all points yielding exactly the obtained travel time difference between source and station. The reflection area can be confined by using adjacent stations.

Knopoff and Mal (1967) proposed a method for the localization of discontinuities, which requires the knowledge of the reflection and transmission coefficients. This method is based on the assumption that a surface wave propagates across a sloping discontinuity with local phase velocity. The amplitudes, however, are influenced by reflection and transmission. For a given arrangment of seismographs the position of a fictitious vertical discontinuity can be calculated. This position is period dependent for a sloping discontinuity and is also influenced by the arrangement of seismographs [see Neunhöfer et al. (1974)]. The accuracy of the determination of the reflecting surface mainly depends on two quantities: on the group velocity in the region under consideration and on the distance between the recording station and the discontinuity. The distance between the stations and the discontinuity should be not too small for the localization of an unknown discontinuity. However, a small distance between the stations right and left of the discontinuity is advantageous for the investigation of the depth penetration and the material contrast of a known discontinuity in order to reduce disturbing influences, which we will speak about later.

Proceeding from the situation mentioned above, the data analysis mainly consists of the measurement of amplitude spectra at both sides of the discontinuity. Subsequently, the quotient is formed and the transmission coefficients are obtained in this way. After separating the reflected waves, the reflection coefficients are obtained similarly. Incidently, Gregersen and Alsop (1974) gave the visual determination of the corresponding amplitude ratios preference over a Fourier analysis. However, as already mentioned, generally the latter has to be applied. Independently of the method of interpretation, it is clear that the propagation of surface waves over a discontinuity is only one of many effects which influence the surface wave amplitudes. According to McGarr (1969a), further influencing factors include source radiation pattern, horizontal refraction, physical and geometrical dissipation, and ambient noise. Furthermore, the uncertainty in the response curves of the seismographs and the properties of the station underground belong to these factors. They all have to be taken into account in investigating the discontinuities. McGarr (1969b) showed that in certain cases the horizontal refraction caused by lateral variations of phase velocity connected with focusing effects can exceed the influence of a vertical discontinuity by an order of magnitude. This effect can be partly repressed by considering simultaneously different epicentral regions.

4.5.3. *Further investigation of discontinuities with surface waves*

4.5.3.1. General remarks

To investigate discontinuities with seismic surface waves, in principle the following procedure is applied. Experimental reflection and transmission coefficients are somehow determined from the seismograms. After that, theoretical coefficients are calculated by using information about the discontinuity obtained by other methods. Subsequently, we have attempted to fit the theoretical coefficients to the measured ones by varying the models right and left of the discontinuity. Of course, this method is not unique as is normal in geophysics. Nevertheless, important additional information about the discontinuities can be obtained in this way. McGarr (1969a) fitted the data on the basis of a rough estimation of the surface amplitude ratio, according to which the energy transported by the fundamental mode is maintained during the passage across the margin. It follows, that the amplitude ratio is proportional to the ratio of energy transports normalized to unit vertical amplitudes at the surfaces right and left of the discontinuity, and it can be changed effectively by variation of precisely this latter ratio. It is especially sensitive to the shear wave velocity and the amplitude-depth dependence of the surface wave.

4.5.3.2. Methods used in seismological literature for the determination of reflection and transmission coefficients of surface waves

4.5.3.2.1. Least-square boundary residual (LQBR) methods

The theoretical background of these methods has already been considered in section 4.4.9., so we want to treat here the seismological applications. For the first time, a LQBR-method was introduced into seismological literature by Alsop (1966) before this method was in fact applied to discontinuities in closed [Davies (1973)] and open dielectric wave-guides [Morishita et al. (1979)]. Alsop (1966) calculated approximate reflection and transmission coefficients of vertically incident Love waves for simple discontinuities: The approximation was that only the discrete Love modes were used in applying the mode-matching procedure, so that the continuity conditions on the interface could be fulfilled in a least squares sense only. This was carried out in such a way that the continuity conditions for the displacements and stresses were combined by applying the orthogonality relation and a functional relationship between the reflection and transmission coefficients followed. The minimum condition for the stress discontinuity on the interface yielded a necessary additional equation for the determination of both the reflection and transmission coefficients. McGarr and Alsop (1967) applied the same procedure for vertically incident Rayleigh waves. The advantage of this method is that it can be applied to situations involving wave-guides of any degree of complexity on either side of the discontinuity as long as the corresponding eigenfunctions are known. A step in elevation can also be taken into account. However, it is impossible to make reliable statements about the accuracy of this method, because the previously mentioned parameter PEUF (see section 4.3.6.) is only partly suitable for this purpose.

Malischewsky (1973) generalized Alsop's method for obliquely incident surface waves. Appearing for the first time, the Rayleigh and Love waves were considered as an uniform eigenfunction system in this paper, and consequently it was possible to take into account their mutual conversion. On this basis, reflection and transmission coefficients were calculated as a function of the angle of incidence and the period as well [see MALISCHEWSKY (1976, 1977)]. Since the rather arbitrary choice of the functional for the variational method (F_1) was unsatisfactory and also led to inconsistencies, it seemed fitting to improve the method by changing the functional (F_2). This was carried out independently of each other by LEVSHIN and YANOVSKAYA (1976) and ITS and YANOVSKAYA (1977, 1979a) on the one hand and by MALISCHEWSKY (1979, 1980) on the other hand. The work of LEVSHIN and YANOVSKAYA (1976) is based on Huygens's principle, where the used sources are equivalent to the stress and displacement discontinuities. For oblique incidence, however, it is more convenient to use the mathematical apparatus of Green's function [ITS and YANOVSKAYA (1979b)]. In each case, the reflection and transmission coefficients determined in this way are equivalent to those which follow from the stationarity condition for the functional F_2. The Green's function used in the paper mentioned above is an approximation only, because it refers to the half-space and consequently neglects reflected waves. In a further paper, ITS and YANOVSKAYA (1982) applied an improved Green's function, but the deviations are essential only for angles of incidence in the over-critical range. The

Table 7. Model parameters for different discontinuity models

Model	Layer	α km/s	β km/s	ϱ g/cm³	Thickness km
MAR1	1		3.2	2.7	4
	2		4.8	3.2	∞
MAR2	1		3.2	2.7	10
	2		4.8	3.2	∞
TAZ1	1	1.95	1.09	1.4	15
	2	5.35	3.15	2.7	∞
TAZ2	1	1.95	1.09	1.4	20
	2	5.35	3.15	2.7	∞
IRI1	1	2.0	0.8	2.0	0.05
	2	2.2	1.0	2.1	0.2
	3	4.4	2.4	2.4	∞
IRI2	1	2.0	0.8	2.0	0.05
	2	3.2	1.6	2.3	0.2
	3	4.4	2.4	2.4	∞

work by Its and Yanovskaya concerning discontinuities was summarized in a recent paper Its and Yanovskaya (1985). You can find a detailed representation of these problems in the monography of Levshin et al. (1986).

4.5.3.2.2. Numerical results on the basis of the LQBR-method

In the following, we present and discuss some model calculations which were carried out with the computer program developed in Leningrad. The model parameters are given in Table 7.

The discontinuity model MAR1/MAR2 is the simplest model of an ocean-continent transition. It was used by Martel (1980) to study the behaviour of

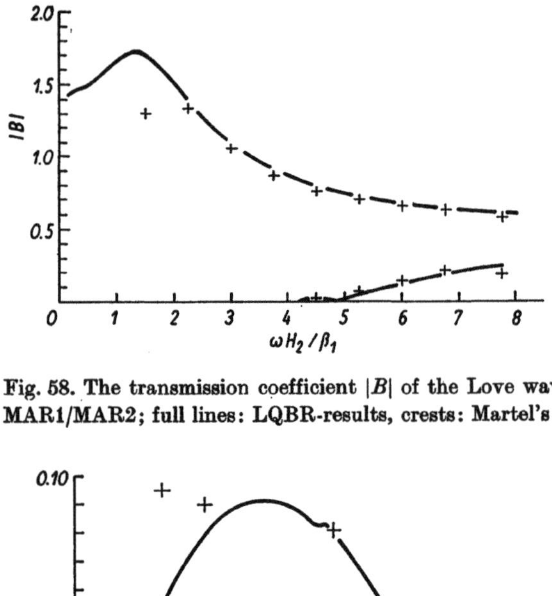

Fig. 58. The transmission coefficient $|B|$ of the Love waves for the model MAR1/MAR2; full lines: LQBR-results, crests: Martel's results

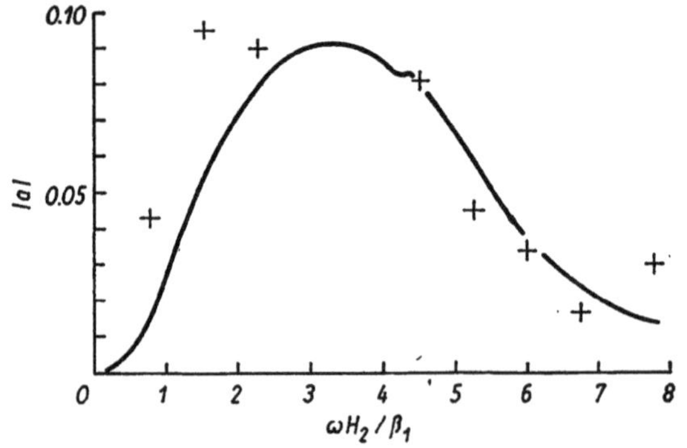

Fig. 59. The reflection coefficient $|a|$ of the Love waves for the model MAR1/MAR2

Love waves at such a step discontinuity using the method of finite elements. In the following two figures, we compare Martel's results with ours. The surface transmission coefficient B for the fundamental and first higher mode of Love waves is shown in Fig. 58 as a function of a dimensionless quantity which is proportional to the circle frequency ω. The trend of the reflection coefficient for the fundamental mode is represented in Fig. 59.

Generally, the coincidence with Martel's values is good, and for the transmission coefficient in the range $\omega H_2/\beta_1 \geqq 2.5$ even very good. The results of GREGERSEN and ALSOP (1974) obtained on the basis of a ray-theoretical method (see section 4.5.3.2.3.) and quoted in MARTEL (1980) are comparable with the other ones.

The propagation of Rayleigh waves across a step discontinuity (propagation from the thin to the thick layer) is studied using the model TAZ1/TAZ2.

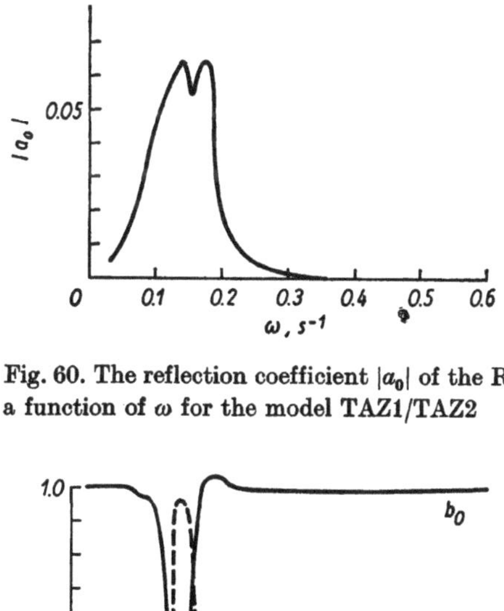

Fig. 60. The reflection coefficient $|a_0|$ of the Rayleigh fundamental mode as a function of ω for the model TAZ1/TAZ2

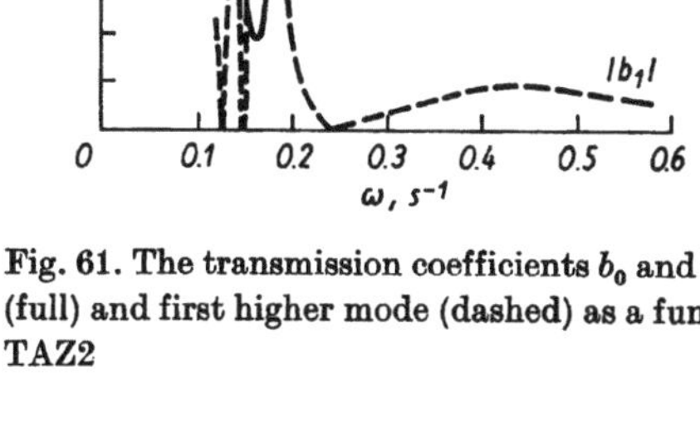

Fig. 61. The transmission coefficients b_0 and b_1 of the Rayleigh fundamental (full) and first higher mode (dashed) as a function of ω for the model TAZ1/TAZ2

The Figs. 60 and 61 show the absolute values of the reflection coefficient a_0 of the fundamental mode and the transmission coefficients b_0 and b_1 of the fundamental and first higher mode respectively.

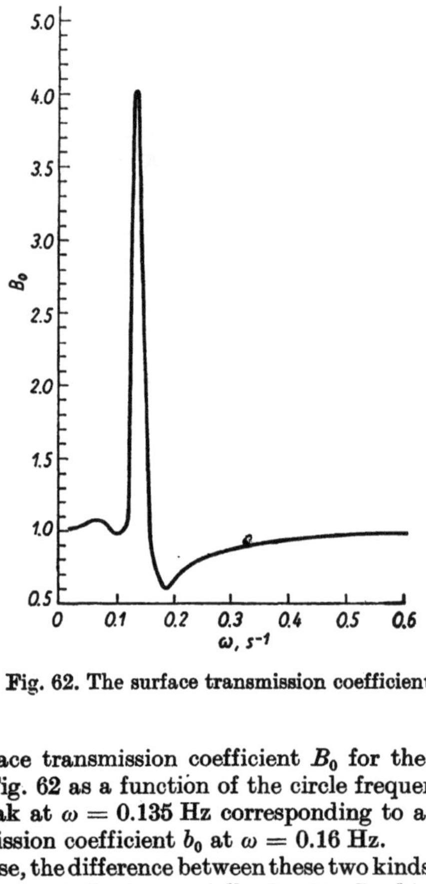

Fig. 62. The surface transmission coefficient B_0 for the model TAZ1/TAZ2

The surface transmission coefficient B_0 for the fundamental mode is presented in Fig. 62 as a function of the circle frequency ω. Note, that there is a distinct peak at $\omega = 0.135$ Hz corresponding to a characteristic minimum of the transmission coefficient b_0 at $\omega = 0.16$ Hz.

In this case, the difference between these two kinds of transmission coefficients mentioned repeatedly is especially drastic. Looking at the normalized eigenfunctions of the Rayleigh waves at the peak frequency $\omega = 0.16$ Hz, it is noted that the amplitude of the fundamental mode for model TAZ2 essentially exceeds that one of model TAZ1. However, the absolute values of the amplitudes of the fundamental mode for TAZ1 and of the first higher mode for TAZ2 have the same order of magnitude in the surface region. So it is understandable, that the transmission coefficient is small for the fundamental mode but great for the first higher mode at this frequency (see Fig. 61). Quoting IRIKURA and KAWANAKA (1980), the discontinuity model TAZ1/TAZ2 (with layer thicknesses H_1 and H_2 reduced by 10^{-6} and an frequency enlarged by 10^6) was the basis for model experiments carried out by Mori and Tazime to verify theoretically predicted spectral ratios for this step discontinuity. Let Y_1 and Y_2 be the spectra for the model with the thinner and the thicker layer respectively, measured at a distance Δ from the source, and let $\bar{Y}$ be the spectrum, if the step discontinuity is located at a distance $\Delta/2$ from the source (see Fig. 63), then the following estimation was obtained by Mori and Tazime:

$$\bar{Y} \approx \sqrt{Y_1 \cdot Y_2}, \tag{4.323}$$

from which an approximated transmission coefficient

$$B \approx \sqrt{Y_1 \cdot Y_2}/Y_1 \tag{4.324}$$

follows, being valid for the direction of propagation TAZ1 → TAZ2.

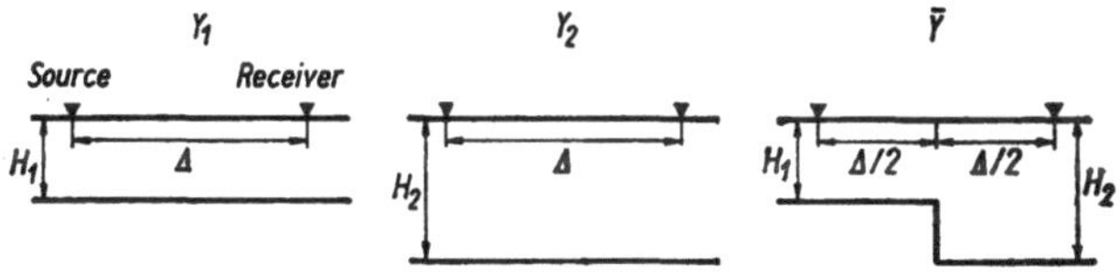

Fig. 63. Geometry for the model according to Mori and Tazime

Instead of the observed spectra Y_1 and Y_2, Irikura and Kawanaka (1980) used spectra which were theoretically obtained by means of Harkrider's formulation for excitation functions of Rayleigh waves in layered media. The semi-analytical transmission coefficients B obtained in this way agree well with the LQBR-approximation (see Fig. 62). According to Irikura and Kawanaka (1980), there is also a good agreement with the method of Alsop et al. (1974). The calculated reflection coefficients are too small to be verified experimentally.

We will likewise study the passage of Rayleigh waves across a discontinuity with the LQBR-method, using the model IRI1/IRI2 (or IRI2/IRI1 for the opposite direction). Irikura and Kawanaka (1980) used this model to simulate a real geological situation in the southern part of the Kyoto basin in Japan, which was important for the interpretation of spectra of microtremors. Fig. 64 shows the transmission coefficients of the fundamental and first higher mode for the direction IRI1 → IRI2 as a function of ω and that for b_0, b_1 and B_0, B_1 as well. The frequency dependence of b_0 is physically plausible: the discontinuity located in the second layer is not "perceived" for very great and very small periods, i.e. b_0 asymptotically reaches the value one. A characteristic minimum is in between, indicating the period range with a maximal influence by the discontinuity.

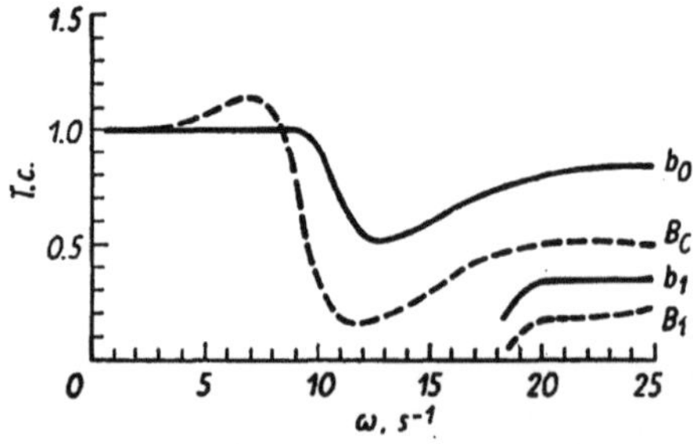

Fig. 64. Transmission coefficients for the model IRI1/IRI2, full lines: "energetic" coefficients, dashed lines: "surface" coefficients

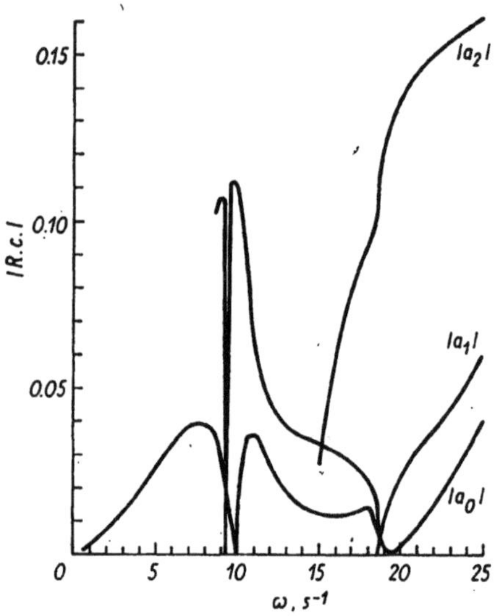

Fig. 65. Reflection coefficients for the model IRI1/IRI2

The curve for B_0 looks like the latter one, in contrast to the model discussed above. It seems to be that such a behaviour is always observed in those cases if the transition takes place from a medium with a lower surface wave velocity to a medium with a higher one. The absolute values of the reflection coefficients belonging to this case are presented in Fig. 65 for the fundamental and the first two higher modes as well.

The model calculations of this section were carried out so that all higher modes occurring in the considered frequency or period range were taken into account.

The following two figures refer to the direction of propagation IRI2 → IRI1. The transmission coefficients b_0, b_1, b_2, B_0, and $|B_1|$ are given in Fig. 66. The curve for b_0 is identical with that of the model IRI1/IRI2 (see Fig. 64). This is a result of the so-called reciprocity, about which we cannot speak here in great detail. For a short time, denote the (normalized) transmission coefficients concerning a mode with mode number α incident from the left with

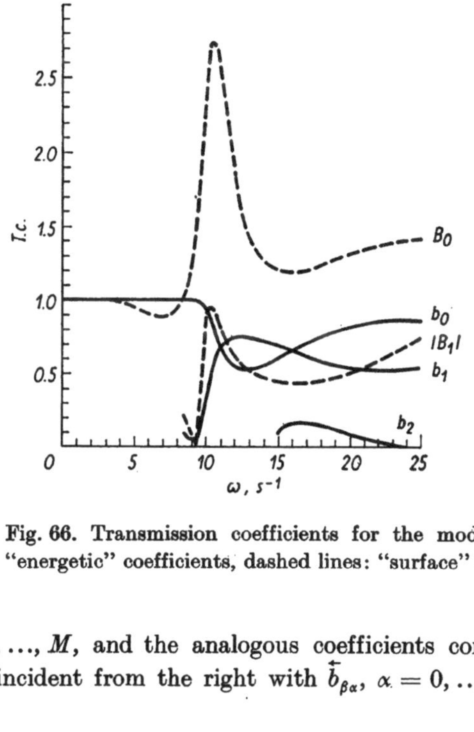

Fig. 66. Transmission coefficients for the model IRI2/IRI1, full lines: "energetic" coefficients, dashed lines: "surface" coefficients

$\vec{b}_{\alpha\beta}$, $\beta = 0, \ldots, M$, and the analogous coefficients concerning a mode with number β incident from the right with $\overleftarrow{b}_{\beta\alpha}$, $\alpha = 0, \ldots, N$. Then, reciprocity means

$$\vec{b}_{\alpha\beta} = \overleftarrow{b}_{\beta\alpha}; \qquad \alpha = 0, \ldots, M, \qquad \beta = 0, \ldots, N. \tag{4.325}$$

According to GREGERSEN (1974), this relation can be obtained in a simple manner from the structure of the differential equations and their solutions which are a superposition of normal modes using the simplifyed assumption that no energy is scattered into body waves. On the basis of a more rigorous treatment, KAZI (1980) showed that this reciprocity is a result of Betti's theorem and does not depend on the used approximation. Betti's theorem is a vector theorem which is equivalent to Green's theorem for scalars and the Laplace's operator [see e.g. AKI and RICHARDS (1980)]. However, KAZI's (1980) proof is valid for the closed wave-guide and Love-kind motion only.

We can establish from Fig. 66 that in the frequency range $\omega \approx 11 \cdots 16\ s^{-1}$, the first higher mode transports more energy in medium 2 than the fundamental one. Furthermore, note that there is a characteristic maximum of the surface transmission coefficient for the fundamental mode B_0 at $\omega \approx 10.5\ s^{-1}$ and a weaker one for the first higher mode at the same place.

Finally, the absolute values of the reflection coefficients are plotted in Fig. 67 for the fundamental mode $|a_0|$ and the first higher mode $|a_1|$ as a function of ω.

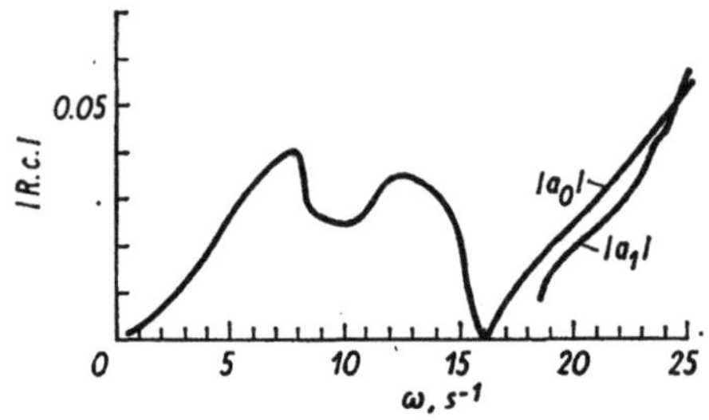

Fig. 67. Reflection coefficients for the model IRI2/IRI1

To some extent, the complicated trend of the reflection coefficients (see also Fig. 65) indicates the uncertainty with which they are determined by the approximation technique.

4.5.3.2.3. Ray-theoretical methods

Ray-optical techniques have been employed extensively for the analysis of high-frequency radiation and scattering problems. Their application to wave-guide problems, however, has been limited. Iskander and Hamid (1975) referred to the importance of these methods concerning the theory of electromagnetic wave-guides. It was pointed out that for various cases the ray-optical method is advantageous, because the solution is less complicated than that obtained using mode-matching techniques.

The essential step in applying ray-theoretical methods is the decomposition of the wave-guide modes into two systems of plane waves crossing each other. To a certain extent, Babich and Molotkov (1966) for example introduced these methods into the theory of seismic surface waves. For the first time, Gregersen and Alsop (1974) calculated approximated reflection and transmission coefficients of Love waves using a ray-theoretical method. The starting-point was the ray-optical decomposition of the surface waves into homogeneous and inhomogeneous body waves. Proceeding from the simplest model "layer over half-space", the Love wave motion is interpreted by up- and down-going homogeneous SH-waves in the layer on the one hand and by inhomogeneous SH-waves in the half-space on the other hand. If these SH-waves strike the discontinuity, the well-known formulas for the reflection and transmission of both homogeneous and inhomogeneous SH-waves can be applied, and in this way a depth-dependent displacement and stress distribution along the interface caused by the incident surface waves follows (see Fig. 68).

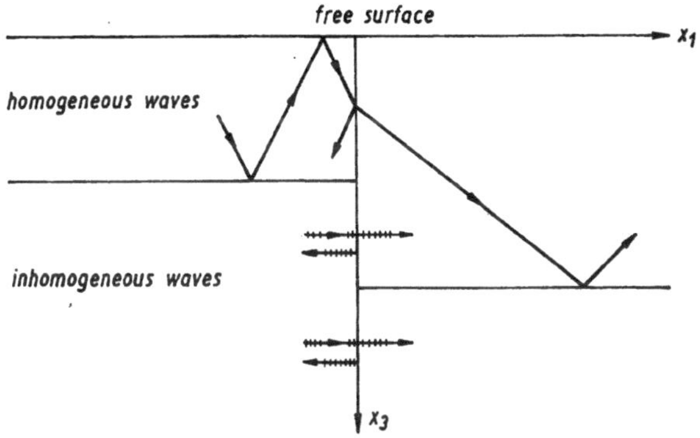

Fig. 68. Decomposition of surface waves at a discontinuity according to GREGERSEN and ALSOP (1974)

Precisely this displacement and stress field generates the reflected and transmitted surface waves at the discontinuity. Using the so-called coupling coefficients, this field can be "projected" into the amplitude-depth functions of the corresponding modes and that part excited by each surface wave mode is obtained. The coupling coefficients can be deduced from HERRERA's (1964) orthogonality relation for elastic surface waves. On the whole, they are identical with the products of the corresponding displacement and stress components integrated over x_3 from 0 up to ∞. This orthogonality relation was generalized by MALISCHEWSKY (1970) in view of the three-dimensional case considered by GREGERSEN and ALSOP (1974). Obviously, by this procedure, the boundary conditions are exactly satisfied on the vertical interface, but not on the free surface and on the horizontal interfaces. This gives rise to diffracted body waves, which are not taken into account by this method. The advantage is the speed of the computation, even for oblique incidence. However, the quality of the approximation can be tested only by comparing with other methods. The same method was also applied to the passage of Rayleigh waves through a vertical discontinuity, and that for normal incidence by ALSOP et al. (1974) and for oblique incidence by CHEN and ALSOP (1979). The process of mutual conversion of Rayleigh and Love waves was investigated by GREGERSEN (1978) using the same method.

WOODHOUSE (1974) developed a kind of ray theory for surface waves in a layered elastic medium in which there are gradual lateral variations in the thicknesses of the layers and in the elastic parameters characterizing each layer. An approximation equivalent to the WKBJ approximation is used, and equations governing the slow variations in amplitude, frequency and wave number in a nearly uniform sinusoidal wave train are deduced. This perturbation procedure takes advantage of the propagator formalism for the unperturbed problem.

SAASTAMOINEN (1984) presented a ray method by which the complex spectra of normal mode surface waves, given a priori, in a local inner product space, living on a vertical cross-section of the wave-guide, can be propagated into another local inner product space over the relevant cross-section. Laterally, the material and the geometric properties of the wave-guide should vary smoothly. The motivation is to create realistic, physical models which, together with the seismic surface wave observations, in a certain network of seismic stations, could reveal further details about the structure of the underlying wave-guide.

4.5.3.2.4. Kazi's method

Kazi and co-workers [see e.g. KAZI (1978a), KAZI (1978b), NIAZY and KAZI (1980), KAZI and NIAZY (1982a), WILLMER and KAZI (1983), and KAZI and NIAZY (1984)] treated the problem of reflection and transmission of Love and also partly of Rayleigh waves at vertical discontinuities, having an increasing degree of complexity, in a theoretically rigorous manner. The method is based on an integral equation formulation together with the application of the Schwinger-Levine variational principle. Let us sum up the main idea of this method for the impedance jump model using the notation developed in this book. Let $\tau(x_3)$ denote the stress components $\sigma_{12\text{res}}^{(1)}$ and $\sigma_{12\text{res}}^{(2)}$ from (4.137) and (4.138) in the "aperture plane" $x_1 = 0$. Then, using the orthogonality relations (3.68), (3.69), and (3.75) and the normalization according to (3.76) and (3.77), the coefficients a_0, b_0, $a(\gamma)$, and $b(\gamma)$ are obtained as

$$1 - a_0 = \frac{\omega}{2\mathrm{i}} \int_0^\infty \tau(x_3)\, U_1^*(x_3)\, \mathrm{d}x_3, \tag{4.326}$$

$$b_0 = \frac{\omega}{2\mathrm{i}} \int_0^\infty \tau(x_3)\, U_2^*(x_3)\, \mathrm{d}x_3, \tag{4.327}$$

$$a(\gamma) = \frac{\mathrm{i}}{\mu} \frac{1}{\sqrt{k_\beta^2 - \gamma^2}} \int_0^\infty \tau(x_3)\, U_1^*(x_3, \gamma)\, \mathrm{d}x_3, \tag{4.328}$$

$$b(\gamma) = -\frac{\mathrm{i}}{\mu} \frac{1}{\sqrt{k_\beta^2 - \gamma^2}} \int_0^\infty \tau(x_3)\, U_2^*(x_3, \gamma)\, \mathrm{d}x_3. \tag{4.329}$$

Substituting these expressions into the continuity condition for the displacement (4.139) and introducing the quantity $\bar{G}(x_3, \bar{x}_3)$, defined by

$$\bar{G}(x_3, \bar{x}_3) = \frac{1}{\mu}\left[\int_0^\infty \frac{U_1^*(x_3, \gamma)\, U_1(x_3, \gamma)}{\sqrt{k_\beta^2 - \gamma^2}}\, \mathrm{d}\gamma + \int_0^\infty \frac{U_2^*(\bar{x}_3, \gamma)\, U_2(x_3, \gamma)}{\sqrt{k_\beta^2 - \gamma^2}}\, \mathrm{d}\gamma\right], \tag{4.330}$$

the following equation is obtained:

$$U_1(x_3)\,(1 + a_0) - U_2(x_3)\, b_0 = \frac{1}{\mathrm{i}} \int_0^\infty \tau(\bar{x}_3)\, \bar{G}(x_3, \bar{x}_3)\, \mathrm{d}\bar{x}_3, \tag{4.331}$$

so that the integral equation formulation is given by (4.326), (4.327), and (4.331). In analogy to the quantum theory, Kazi now introduces a scattering matrix, which in principle contains the required complex reflection and transmission coefficients. However, for the simple model discussed here, the introduction of a scattering matrix has no sense. Up to this stage, the formulation of the problem is exact, but unfortunately it is not possible to obtain the exact solution and therefore approximate ones have to be constructed. According to Kazi (1978), the first way to do this is based upon neglecting the modes arising from the continuous spectrum. This has already been done by us in great detail, so that we will treat now the second method which is connected with the application of the Schwinger-Levine variational principle to improve the earlier approximation in such a way as to incorporate the effects of the body waves indirectly. To do this, we will follow the representation of Collin (1960). Substituting (4.326) and (4.327) into (4.331) and defining

$$G(x_3, \bar{x}_3) = \bar{G}(x_3, \bar{x}_3) + \frac{\omega}{2}\, U_2(x_3)\, U_2^*(x_3) \tag{4.332}$$

and

$$Y = \frac{1 + a_0}{1 - a_0} \tag{4.331}$$

we obtain

$$Y\omega U_1(x_3) \int_0^\infty \tau(\bar{x}_3)\, U_1^*(\bar{x}_3)\, \mathrm{d}\bar{x}_3 = \int_0^\infty \tau(\bar{x}_3)\, G(x_3, \bar{x}_3)\, \mathrm{d}\bar{x}_3, \tag{4.334}$$

where Y is equivalent to the so-called input impedance in electromagnetic theory. According to Collin (1960), the application of the Schwinger-Levine variational principle yields the following variational expression for Y:

$$Y = \frac{1}{\omega} \frac{\int\limits_0^\infty \int\limits_0^\infty G(x_3, \bar{x}_3)\, \tau(x_3)\, \tau(\bar{x}_3)\, dx_3\, d\bar{x}_3}{\left[\int\limits_0^\infty \tau(x_3)\, U_1(x_3)\, dx_3\right]^2}, \tag{4.335}$$

which has the property that Y is stationary for small arbitrary variations in the stress distribution $\tau(x_3)$ about its correct value. That means that a first-order approximation to the stress distribution will yield a second-order approximation to the input impedance. Also the expression for Y is homogeneous in the sense that it does not depend on the amplitude of the stress distribution, but only on its functional form. For a suitable approximation of the "aperture field" $\tau(x_3)$, we may take a finite series of the eigenfunctions of the left domain, say

$$\tau(x_3) \overset{!}{=} \sum_{\alpha=0}^{N_{max}} d_\alpha{}^\alpha U_1(x_3). \tag{4.336}$$

We must now require that the coefficients d_α make Y stationary. This implies

$$\frac{\partial Y}{\partial d_\alpha} = 0, \qquad \alpha = 0, \ldots, N_{max}, \tag{4.337}$$

and gives the best possible solution for Y that can be obtained with the assumed approximation for τ. In our simple case, there is one discrete eigenfunction only, namely $U_1(x_3)$, and we obtain directly from (4.335)

$$Y = \frac{1}{\omega} \frac{\int\limits_0^\infty \int\limits_0^\infty G(x_3, \bar{x}_3)\, U_1(x_3)\, U_1'(\bar{x}_3)\, dx_3\, d\bar{x}_3}{\left[\int\limits_0^\infty U_1^2(x_3)\, dx_3\right]^2}. \tag{4.338}$$

This equation can be simplified considerably by taking into account (3.68), (3.69), and (3.76). Substituting (4.330) and (4.332) into (4.338) and using the notation (4.148) we obtain after some algebra

$$Y = \frac{\mu^2 k_1^2 \omega^2}{8} P^2 + \frac{\mu k_1^2 \omega}{4} \int\limits_0^\infty \frac{R(\gamma)\, R^*(\gamma)}{\sqrt{k_\beta^2 - \gamma^2}}\, d\gamma \tag{4.339}$$

and finally by introducing the expressions (4.164) and (4.166) for the scalar products

$$Y = 2 \frac{k_1}{k_2} \frac{\varepsilon_1 \varepsilon_2}{(\varepsilon_1 + \varepsilon_2)^2} + \frac{2\mu^3 k_1 \varepsilon_1 (\Delta\varepsilon)^2}{\pi} \int\limits_0^\infty \frac{\gamma^2\, d\gamma}{(\varepsilon_1^2 + \gamma^2\mu^2)^2\, (\varepsilon_2^2 + \gamma^2\mu^2)\, \sqrt{k_\beta^2 - \gamma^2}}. \tag{4.340}$$

Using the substitution (4.208) the integral can be transformed into a rational one, however the evaluation of the latter is much more complicated than in the case (4.209). Nevertheless, in principle the integral could be evaluated exactly. For the more complex models used by Kazi, in each case the integrals which occurred had to be calculated numerically. After evaluating the integral in (4.340) it would be interesting to compare the results with the exact ones according to Wiener-Hopf to estimate the quality of this approximation method. But this cannot be carried out here. However, in the author's opinion the quality of Kazi's method should be comparable with that of the second-order theory developed in section 4.4.6.2.

In the following, let us sum up the essential features of Kazi's numerical results. It was shown that the continuous part of the spectrum of the Love wave operator, which is associated with the generation of body waves, can be very important if there is a large contrast in the elastic constants of the media on the two sides of the discontinuity. For the latter case, the most important difference is connected with spike-like features in the transmission coefficient if the body wave contribution has been included. These spikes are associated with the generation of the higher modes above the fundamental one. Increasing the contrast also causes an increase in the sharpness of those spike-like features. The body wave contribution is quite sensitive to the generation of higher modes. Neglecting the body wave contribution leads to an over-estimation of the transmission coefficient that can be quite substantial when there is a large contrast in the elastic parameters on the two sides of the discontinuity. Likewise, neglecting the higher modes leads to a significant over-estimation of the transmission coefficients if the fundamental mode is incident. All reflection coefficients exhibit somewhat similar sensitivities to the higher modes and to the body-waves effect. For Kazi's models, the effect of the higher modes and body waves diminishes with increasing frequency and is negligible for extremely low frequencies.

Using the Schwinger-Levine method HUDSON (1977) also considered the transmission of Love waves through a laterally varying structure. His model consists of a plane-layered structure on the left joined to a (possibly different) plane-layered structure on the right by a region within which the surface elevation and the elastic properties may change from the lateral uniformity outside. Stationary functionals of the displacement function for transmission and reflection ratios of the separate modes and the far-field body-wave radiation were constructed. Certainly, the model is more complex than Kazi's. However, as a result of numerical difficulties, explicit expressions for the reflection and transmission coefficients have not been given.

4.5.3.2.5. Bukchin's method

Using a half-analytical method on the basis of a modified mode-matching procedure, BUKCHIN and LEVSHIN (1980) succeeded in determining the reflection and transmission coefficients as well as the energy converted into body waves for perpendicularly incident Love waves at a step discontinuity. The crucial step was the application of a locally-bounded version of Schwarz's technique [see ZAVADSKY (1972)]. Let us sum up the main ideas using the impedance jump model. In the first instance, the procedure is similar to Kazi, except that, instead of $\tau(x_3)$, now the resulting displacements $u_{\text{res}}^{(1)}$ and $u_{\text{res}}^{(2)}$ [see (4.135) and (4.136)] are used. Let $v(x_3)$ denote the latter in the "aperture plane" $x_1 = 0$. Furthermore, let us formulate the problem for the right and left quarter-spaces separately. For the right one, $u_{\text{res}}^{(2)}(x_1, x_3)$ should satisfy the equation

$$\Delta u_{\text{res}}^{(2)} + k_\beta^2 u_{\text{res}}^{(2)} = 0 \tag{4.341}$$

and the boundary conditions

$$u_{\text{res}}^{(2)}(0, x_3) = v(x_3), \tag{4.342}$$

$$\mu \frac{\partial u_{\text{res}}^{(2)}}{\partial x_3}(x_1, 0) + \varepsilon_2 u_{\text{res}}(x_1, 0) = 0, \qquad x_1 > 0, \tag{4.343}$$

together with convenient conditions at infinity. The solution for the left quarter-space may be sought in the form

$$u_{\text{res}}^{(1)}(x_1, x_3) = u_{\text{inc}}(x_1, x_3) - u_{\text{inc}}(-x_1, x_3) + u_1(x_1, x_3), \tag{4.344}$$

where u_{inc} represents the incident Love-kind wave also satisfying a Helmholtz equation of the form (4.341). Then $u_1(x_1, x_3)$ is a solution of a problem of the same kind as for the right quarter-space, but with slightly different boundary conditions

$$u_{\text{res}}^{(1)}(0, x_3) = u_1(0, x_3) = v(x_3), \tag{4.345}$$

$$\mu \frac{\partial u_1(x_1, 0)}{\partial x_3} + \varepsilon_1 u_1(x_1, 0) = 0, \qquad x_1 < 0. \tag{4.346}$$

Using the orthonormality relations, the expansion coefficients needed by us can be expressed by the "aperture displacement" $v(x_3)$ analogous to (4.326) to (4.329) as follows:

$$a_0 = \frac{\mu k_1 \omega}{2} \int_0^\infty v(x_3)\, U_1^*(x_3)\, \mathrm{d}x_3, \tag{4.347}$$

$$b_0 = \frac{\mu k_2 \omega}{2} \int_0^\infty v(x_3)\, U_2^*(x_3)\, \mathrm{d}x_3, \tag{4.348}$$

$$a(\gamma) = \int_0^\infty v(x_3)\, U_1^*(x_3, \gamma)\, \mathrm{d}x_3, \tag{4.349}$$

$$b(\gamma) = \int_0^\infty v(x_3)\, U_2^*(x_3, \gamma)\, \mathrm{d}x_3. \tag{4.350}$$

For finding $v(x_3)$, Schwarz's technique is now applied. Firstly, $v(x_3)$ is represented by means of displacements $u_{\text{res}}^{(1)}(-\bar{\varepsilon}, x_3)$ and $u_{\text{res}}^{(2)}(\bar{\varepsilon}, x_3)$ for the planes $x_1 = -\bar{\varepsilon}$ and $x_1 = \bar{\varepsilon}$, where $\bar{\varepsilon}$ is a small real quantity. To do this, the condition of continuity of stresses at the plane $x_1 = 0$ is replaced by its finite difference analogue:

$$\mu \frac{u_{\text{res}}^{(1)}(0, x_3) - u_{\text{res}}^{(1)}(-\bar{\varepsilon}, x_3)}{\bar{\varepsilon}} = \mu \frac{u_{\text{res}}^{(2)}(\bar{\varepsilon}, x_3) - u_{\text{res}}^{(2)}(0, x_3)}{\bar{\varepsilon}} + O(\bar{\varepsilon}). \tag{4.351}$$

Neglecting small quantities $O(\varepsilon^2)$ we obtain

$$v(x_3) = \frac{u_{\text{res}}^{(1)}(-\bar{\varepsilon}, x_3) + u_{\text{res}}^{(2)}(\bar{\varepsilon}, x_3)}{2}. \tag{4.352}$$

Substituting $u_{\text{res}}^{(1)}(-\bar{\varepsilon}, x_3)$ and $u_{\text{res}}^{(2)}(\bar{\varepsilon}, x_3)$ from (4.135), (4.136), and (4.344) into this equation and using (4.347)–(4.350) the following integral equation for $v(x_3)$ is obtained:

$$\begin{aligned} v(x_3) = \frac{1}{2} \Bigg[& u_{\text{inc}}(-\bar{\varepsilon}, x_3) - u_{\text{inc}}(\bar{\varepsilon}, x_3) \\ & + \frac{\mu k_1 \omega}{2} U_1(x_3) \, e^{i k_1 \bar{\varepsilon}} \int_0^\infty v(x_3) \, U_1^*(x_3) \, dx_3 \\ & + \int_0^\infty \int_0^\infty v(x_3) \, U_1(x_3, \gamma) \, U_1^*(x_3, \gamma) \, e^{i\bar{\varepsilon}\sqrt{k_\beta^2 - \gamma^2}} \, dx_3 \, d\gamma \\ & + \frac{\mu k_2 \omega}{2} U_2(x_3) \, e^{i k_2 \bar{\varepsilon}} \int_0^\infty v(x_3) \, U_2^*(x_3) \, dx_3 \\ & + \int_0^\infty \int_0^\infty v(x_3) \, U_2(x_3, \gamma) \, U_2^*(x_3, \gamma) \, e^{i\bar{\varepsilon}\sqrt{k_\beta^2 - \gamma^2}} \, dx_3 \, d\gamma \Bigg]. \end{aligned} \tag{4.353}$$

It may be shown that the solution of (4.353) differs from the exact solution $v(x_3)$ by a small quantity of order $O(\bar{\varepsilon})$. This may also be justified for complex $\bar{\varepsilon}$ if $\operatorname{Im} \bar{\varepsilon} > 0$. The integral equation (4.353) is solved by successive iterations, which converge for complex $\bar{\varepsilon}$. Having found $v(x_3)$ the reflection and transmission coefficients can be determined using (4.347)–(4.350). The numerical results obtained by Bukchin and Levshin (1980) aggree well with the approximate Green's function technique according to Its and Yanovskaya (1979b).

Furthermore it was found that for the sharp discontinuity under consideration the amplitudes of the reflected waves are relatively small and decrease with increasing period. Finally, non-stationary seismograms have been calculated. It is interesting from a practical point of view to estimate the errors introduced by interference with body waves in surface wave velocity measurements. This was done and the contribution of body waves to the general solution was found to be quite significant, and it may produce distortions in surface wave velocity measurements near the discontinuity. Notwithstanding the fact that the integrals in (4.353) can be estimated efficiently by using Fast Fourier Transforms, the determination of $v(x_3)$ is cumbersome and needs a great amount of calculation. That is why the method can hardly be applied to more complicated models.

4.5.3.2.6. The method of coupled modes after Kennett

The method of coupled modes developed in 4.2.2. was used for seimological purpose by KENNETT (1984). The starting point is given by the well-known fact that the L_g phase consists of a superposition of many higher surface wave modes. At 1 Hz about 15 are important and the number increases rapidly with increasing frequency. The phase is sensitive to variations in crustal structure and observed waveforms can be very complex. The approach was introduced by considering a 2-D situation in which the elastic parameters of the crustal structure are independent of the co-ordinate x_2, and a similar theory as described in 4.2.2.1. is developed. Neglecting body wave interactions the attention is restricted to the modal part of the wave field. The latter one can be expressed locally as a sum of modes travelling in both the positive and negative x_1 directions:

$$\Psi_{\text{mod}}(x_1, x_3) = \sum_{\alpha=1}^{N_{\max}} [C_\alpha^{(+)}(x_1)\, {}^\alpha\Psi^{(+)}(x_1, x_3) + C_\alpha^{(-)}(x_1)\, {}^\alpha\Psi^{(-)}(x_1, x_3)]. \tag{4.354}$$

Concerning ${}^\alpha\Psi^{(+)}$ and ${}^\alpha\Psi^{(-)}$ note, that the reference medium can either have constant properties or we can work in terms of local normal modes. If the model consists of a region of heterogeneity superimposed on an otherwise stratified medium, we work in terms of the modes of the reference stratification. When, however, the model is such that there is a slowly varying change in the large-scale structure, we work in terms of a superposition of local modes. In order to study the development of the wave train in the x_1-direction, an evolution equation (4.34) for the coefficients $C_\alpha^{(\pm)}$ is derived in terms of the perturbations, which constitutes a set of $2N_{\max}$ coupled differential equations of the first order, which have to be solved subject to constraints on the form of the seismic wavefield at the limits of the heterogeneous region. Supposing that the region of departure from stratification is isolated between the positions

x_L and x_R (Fig. 69) and considering a surface wave incident from the left, this fixes ${}^\alpha C^{(+)}(x_L)$ and for $x_1 > x_R$, ${}^\alpha C^{(-)}(x_1) \equiv 0$ must hold. This imposes a two-point boundary value problem for the $2N_{\max}$ coupled equations mentioned above, since the information we seek is the reflected field determined by ${}^\alpha C^{(-)}(x_L)$ and the transmitted field determined by ${}^\alpha C^{(+)}(x_R)$. For a large number of complex equations this poses an awkward numerical problem. In formulas (4.35)–(4.40), we have demonstrated how this problem can be avoided for weak perturbations. Following KENNETT (1984), if the heterogeneity is not required to be small, an alternative procedure is to concentrate directly on the reflection and transmission aspects of the problem. For simpli-

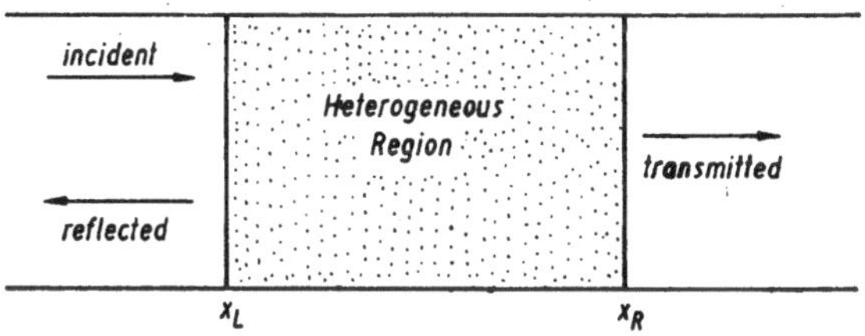

Fig. 69. Configuration of the heterogeneous zone

city, let us begin with a reference medium in a "single mode state" (i.e. constant elastic parameters and impedance surface). Then we have only an incident, a reflected, and a transmitted mode, and the coupled system (4.34) is simply given by

$$\frac{\mathrm{d}C^{(+)}}{\mathrm{d}x_1} = C^{(+)}M_{++} + C^{(-)}M_{+-}\,\mathrm{e}^{-2\mathrm{i}kx_1}, \tag{4.355}$$

$$\frac{\mathrm{d}C^{(-)}}{\mathrm{d}x_1} = C^{(+)}M_{-+}\,\mathrm{e}^{2\mathrm{i}kx_1} + C^{(-)}M_{--}, \tag{4.356}$$

or with the notation of KENNETT (1984)

$$\frac{\mathrm{d}}{\mathrm{d}x_1}\begin{pmatrix} C^{(+)} \\ C^{(-)} \end{pmatrix} = \begin{pmatrix} B^{++} & B^{-+} \\ B^{+-} & B^{--} \end{pmatrix}\begin{pmatrix} C^{(+)} \\ C^{(-)} \end{pmatrix}, \tag{4.357}$$

where

$$\begin{aligned} B^{++} &= M_{++}, & B^{-+} &= M_{+-}\,\mathrm{e}^{-2\mathrm{i}kx_1}, \\ B^{+-} &= M_{-+}\,\mathrm{e}^{2\mathrm{i}kx_1}, & B^{--} &= M_{--}. \end{aligned} \tag{4.358}$$

For a surface wavetrain incident from the left at x_L, we define a transmission matrix $T^{++}(x_\mathrm{R}, x_\mathrm{L})$ (which is the transmission coefficient in the single mode case) connecting the $C^{(+)}$ fields at x_R and x_L, and a reflection matrix $R^{-+}(x_\mathrm{R}, x_\mathrm{L})$

(which is the reflection coefficient in the single mode case) connecting the left and rightward travelling wave fields at x_L:

$$C^{(+)}(x_R) = T^{++}(x_R, x_L)\, C^{(+)}(x_L), \tag{4.359}$$

$$C^{(-)}(x_L) = R^{-+}(x_R, x_L)\, C^{(+)}(x_L). \tag{4.360}$$

Now, by adapting the invariant imbedding technique it is possible to find a set of non-linear differential equations of the Ricatti type for the two $N_{max} \times N_{max}$ matrices R^{-+}, T^{++} as a function of x_L with x_R fixed. This corresponds to considering a sequence of problems with the region of heterogeneity progressively extended towards the left, away from x_R. The equations for R^{-+}, T^{++} are derived by differentiating (4.359) and (4.360) with respect to x_L:

$$\frac{\partial}{\partial x_L} C^{(+)}(x_R) = \frac{\partial}{\partial x_L} T^{++} \cdot C^{(+)}(x_L) + T^{++} \frac{\partial}{\partial x_L} C^{(+)}(x_L) = 0, \tag{4.361}$$

$$\frac{\partial}{\partial x_L} C^{(-)}(x_L) = \frac{\partial}{\partial x_L} R^{-+} \cdot C^{(+)}(x_L) + R^{-+} \frac{\partial}{\partial x_L} C^{(+)}(x_L). \tag{4.362}$$

The derivatives of the coefficient vectors can be found from (4.357) in terms of $C^{(+)}(x_L)$ alone, by substituting for $C^{(-)}(x_L)$ from (4.360), and after introducing these derivatives into (4.361) and (4.362), the following coupled Ricatti equations for R^{-+}, T^{++} are obtained:

$$\frac{\partial}{\partial x_L} R^{-+} + R^{-+}B^{++} - B^{--}R^{-+} + R^{-+}B^{-+}R^{-+} - B^{+-} = 0, \tag{4.363}$$

$$\frac{\partial}{\partial x_L} T^{++} + T^{++}B^{++} + T^{++}B^{-+}R^{-+} = 0. \tag{4.364}$$

For $N_{max} > 1$, all quantities have to be interpreted as matrices with the corresponding matrix multiplications. These equations are integrated starting at x_R and move to the left with the initial conditions that, in the absence of heterogeneity, there is complete transmission and no reflection, i.e.

$$T^{++}(x_R, x_R) = 1, \qquad R^{-+}(x_R, x_R) = 0. \tag{4.365}$$

This set of two $N_{max} \times N_{max}$ Ricatti equations replaces the original $2N_{max}$ coupled first-order equations, and transforms a two-point boundary value problem into an initial value problem which is much easier to solve numerically. According to KENNETT (1984) equation (4.364) gives us a means of assessing the circumstances in which we can neglect the reflected waves and concentrate instead on the interconversions between modes in transmission. If we can neglect the term $T^{++}B^{-+}R^{-+}$ we have a simple first-order equation for the matrix T^{++}

$$\frac{\partial}{\partial x_{\mathrm{L}}} T^{++} + T^{++}B^{++} \approx 0, \tag{4.366}$$

which depends entirely on the coupling coefficients between forward travelling modes. For such a weak heterogeneity, (4.366) provides a rapid method of calculating the transmitted field. However, in most circumstances it would be dangerous to assume that transmission effects are dominant without a prior calculation including reflection.

The method was demonstrated by Kennett (1984) for 1 Hz $\mathrm{L_g}$- and $\mathrm{S_n}$-waves in a multilayered model for which there are 19 coupled modes. Taking a real crustal model as the reference stratification, the perturbation was generated by inserting rectangular blocks of uniform material, and the corresponding reflection and transmission matrices were calculated. Even for this simple heterogeneity, there is a comparatively strong coupling between forward and backward travelling modes including higher ones. The modal amplitudes are given as a function of the width of the heterogeneity, and we observe that the corresponding coefficients reach constant values for increasing width. This is contrary to the familiar phenomenon of reflection and transmission of electromagnetic waves through a layer, which is characterized by a sequence of maxima and minima ($\lambda/4$ layer!). But the most surprising result is the magnitude of certain reflection coefficients (more than 0.4). Generally, such high values are not obtained for a single vertical discontinuity using the LQBR-method.

The approach can be extended to the case of 3-D heterogeneity superimposed on a stratified medium. Note that for a surface wave train travelling in the x_1-direction in a general structure we need to allow for coupling of Love and Rayleigh waves. As already mentioned, the idea of combining Rayleigh and Love waves to form a homogeneous eigenfunction system originates from Malischewsky (1973). Using such a "super-mode" system, the coupling of Rayleigh and Love waves follows naturally from this perturbation theory. The cross-coupling terms Rayleigh → Love and vice versa are given explicitely by Kennett (1984). Furthermore, it is pointed out that in the presence of a transverse parameter gradient, there will be a net transfer of energy from Rayleigh to Love waves.

4.5.3.2.7. Further approximation techniques

Obviously, pure numerical methods and model experiments play an important role in investigating laterally inhomogeneous structures with surface waves. Some references can be found in the introduction. However, we cannot go into details. Here, let us first briefly refer to a heuristical method which, despite its simplicity, yielded good approximations for the reflection coefficient of Love-kind channel waves in coal seams with discontinuities. For the configuration represented in Fig. 70, KORN and STÖCKL (1982) assume approximately that only that part of the wave propagating inside the layer is reflected at the discontinuity with the reflection coefficient r_0 for plane SH waves.

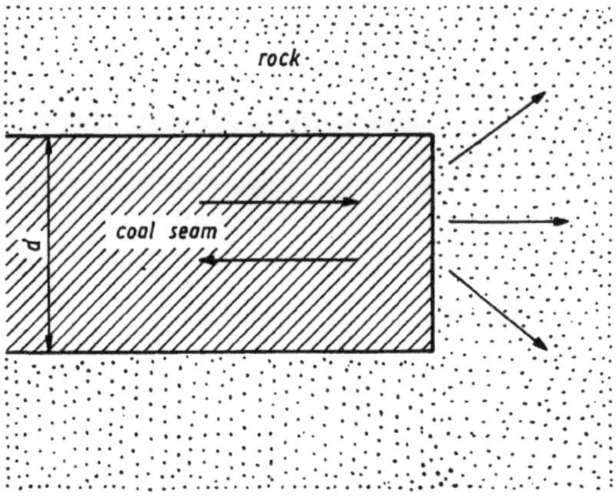

Fig. 70. Reflection of channel waves at a coal seam discontinuity

Moreover, the amplitudes in the neighbouring rock are supposed to be unaffected by the discontinuity. Under this assumption, an approximation for the reflection coefficient r_{th} of the channel waves can be written as

$$r_{th}(\omega) = r_0 \cdot \int_{-d/2}^{+d/2} A(z)\,dz \bigg/ \int_{-\infty}^{\infty} A(z)\,dz, \tag{4.367}$$

where d is the seam thickness, and $A(z)$ is the amplitude-depth distribution of the channel waves. This reflection coefficient agrees very well with the numerically obtained result over the whole range of frequencies. From this we may conclude that the amplitude of the reflected Love wave depends mainly on the amplitude partition of the incident wave between rock and coal, and that other effects, such as diffractions at the seam corners, are not important in this case.

Another attempt to solve the scattering problem of Rayleigh waves at a vertical boundary originates from MOMOI (1981). The method of this approach is such that integral equations are derived by use of the Fourier transform technique and they are solved by deforming their integration path along which the integrals vary smoothly in magnitude. The energy fluxes of the transmitted and reflected Rayleigh waves, the scattered body waves and the Stoneley wave (if it exists) are numerically computed. So Momoi succeeded in solving a very complicated scattering problem with an accuracy which is, according to MOMOI (1981), sufficiently high for the purpose. On the other hand, we cannot evaluate this work here, because it is not easy to comprehend the philosophy. MOMOI (1981) found the following common features:

a) the energy flux of the transmitted Rayleigh waves is the most noteworthy among all the partitioned waves,
b) the forward and backward transmissions of Rayleigh waves in the same model are almost the same in amount,
c) the scattered P waves propagate along the boundaries as if they were trapped there, while the scattered S waves vanish there,
d) in the curves of energy density showing the directivity of the scattered body waves, small fluctuations of sharp upheavals and depressions appear.

Furthermore, an approximate evaluation of the energy flux of the transmitted Rayleigh wave was carried out by assuming that the overlapped area of two energy profiles (if superposed) of Rayleigh waves in the two domains, only contributes to the energy of the transmitted Rayleigh wave. The results obtained in this way are not so bad.

Without going into details we state here that there are also attemps to calculate reflection and transmission coefficients of surface waves for sloping and curved discontinuities which have considerable importance in seismology. For example, MALISCHEWSKY (1974) generalized Alsop's method for curved discontinuities, and using a Green's function technique LUTIKOV (1979) determined reflection and transmission coefficients of Rayleigh waves for a sloping discontinuity as a function of the slope angle.

4.5.3.3. The penetration in depth of discontinuities

The statement about the penetration in depth of lateral perturbations belongs to those which can be obtained directly from the data after a convenient processing without solving the complicated boundary value problem concerning the propagation of surface waves in laterally disturbed media. Let $S_B(\omega)/S_A(\omega)$ be the ratio of the amplitude spectra in the points A and B for propagation in the direction $A \to B$ and $\tilde{S}_A(\omega)/\tilde{S}_B(\omega)$ for the opposite direction $B \to A$. Then, quoting SICHARULIDZE (1978) and neglecting the geometrical and physical attenuation, the following relation

$$K(\omega) = \frac{S_B(\omega)}{S_A(\omega)} \cdot \frac{\tilde{S}_A(\omega)}{\tilde{S}_B(\omega)} \approx 1 \tag{4.368}$$

should hold for an undisturbed wave path. However, if there is any interface between A and B, $K(\omega)$ will be less than 1 due to the losses caused by reflection and scattering. The quantity $1 - K(\omega)$ is a measure for the loss of energy during the passage of the waves through the interface. By "extracting" this quantity carefully from the seismograms we can make at least qualitative statements concerning the penetration in depth of discontinuities between A and B. Let H_T be the penetration in depth of a discontinuity, then a surface wave with a wave length λ, $\lambda \gg H_T$, will not "notice" the discontinuity and consequently, $1 - K(\omega)$ goes to zero for long periods. For short periods, i.e. $\lambda \ll H_T$, $1 - K(\omega)$ should converge to a value other than zero and depending on the elastic parameters at the interface only. We can proceed on the assumption that the period range resulting in a considerable decrease of $1 - K(\omega)$ corresponds to those wave lengths which are comparable with the penetration in depth of the discontinuity. Let us demonstrate this more thoroughly using the following model. Let a layer with a vertical discontinuity be deposited on a half-space, and the thickness H_T of the layer is identical with the penetration in pepth of the discontinuity (Fig. 71).

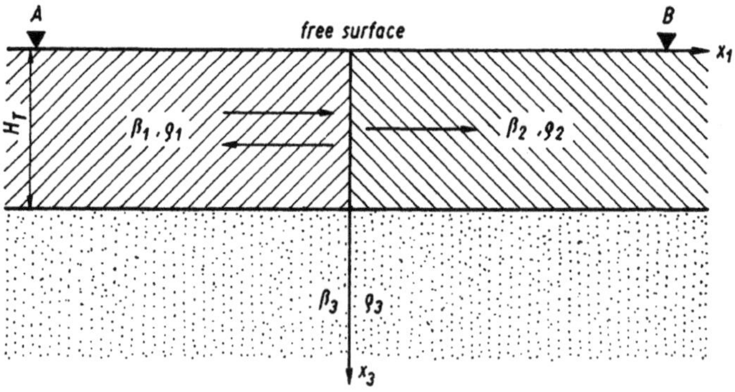

Fig. 71. A simple model of a vertical discontinuity

Using a 2-D configuration we consider Love waves which propagate between the stations A and B. The ratios $S_B(\omega)/S_A(\omega)$ and $\tilde{S}_A(\omega)/\tilde{S}_B(\omega)$ can be identified with the surface transmission coefficients T_{AB} and T_{BA}, respectively, and the latter ones are determined neglecting higher modes with the improved version of Alsop's method [see MALISCHEWSKY (1979)]. The model parameters are given in Table 8.

For the models DISC1 up to DISC4, the limit periods for the 1. higher mode are 5, 3.6, 2.5, and 1.8 s. In Fig. 72, the reflection coefficient a and the energy transmission coefficient b is represented versus period T for the discontinuity models DISC1/DISC2 and DISC3/DISC4, and similarly the surface transmission coefficient $B(= T_{AB})$ in Fig. 73.

The quantity PEUF as a function of the period is given in Fig. 74.

In the short period range, the reflection coefficients reach the limit 0.0769, and the transmission coefficients B 0.9231 corresponding to the reflection

Table 8. Model parameters for simple vertical discontinuities

Model	Layer	β km/s	ϱ g/cm³	Thickness km
DISC1	1	3.0	2.6	10
	2	4.5	3.3	∞
DISC2	1	3.5	2.6	10
	2	4.5	3.3	∞
DISC3	1	3.0	2.6	5
	2	4.5	3.3	∞
DISC4	1	3.5	2.6	5
	2	4.5	3.3	∞

and transmission coefficients for plane waves, in the case that layer 1 is extended up to infinity. This behaviour agrees with the calculations of ITTIPIBOON and HAMID (1981). However, the maximum of the reflection coefficient found by these authors and also by PUDENSI and FERREIRA (1982) in an inter-

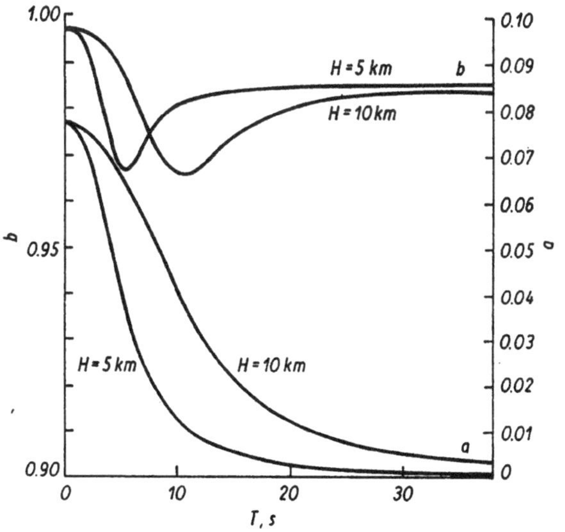

Fig. 72. The coefficients a and b versus period for the models DISC1/DISC2 and DISC3/DISC4

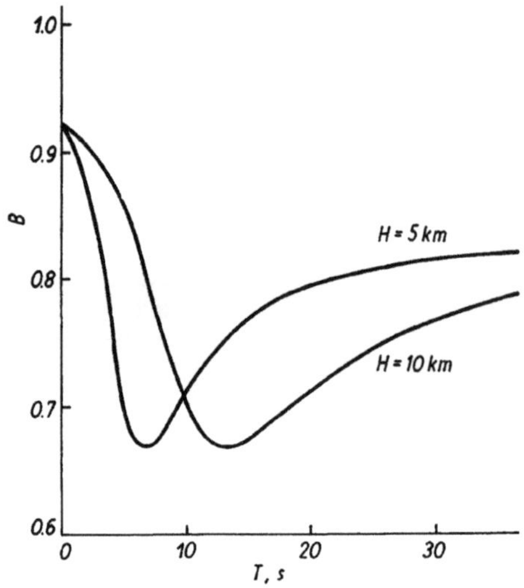

Fig. 73. The coefficient $B(=T_{AB})$ versus period for the models DISC1/DISC2 and DISC3/DISC4

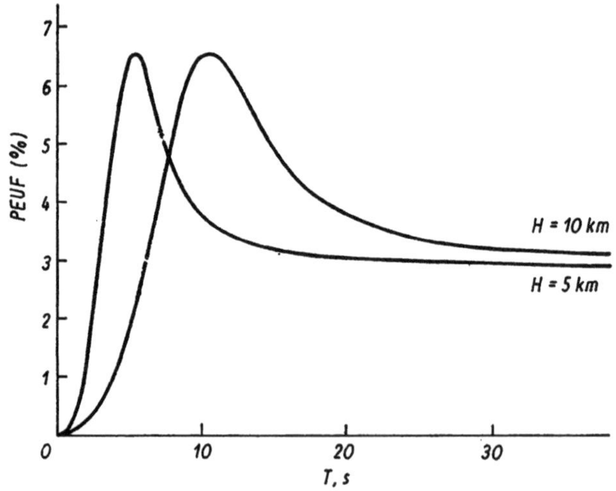

Fig. 74. PEUF versus period for the models DISC1/DISC2 and DISC3/DISC4

mediate period range cannot be verified here. We have also calculated that part of the energy which is transported in the layer by the fundamental mode. As expected, it goes to zero for $T \to \infty$, and it is less than 10% for all models in the period range $T > 27$ s. Consequently, using this fact, the small values of the reflection coefficients in this range may be explained in agreement with STEGEMAN et al. (1981). The curves for the transmission coefficients b are characterized by minima at the periods $\hat{T} \approx 10.5$ s for DISC1/DISC2 and $\hat{T} \approx 5.2$ s for DISC3/DISC4. Denoting the wave length of Love waves averaged over the media right and left of the discontinuity with $\bar{\lambda}_L$, then the penetration into depth of the discontinuity is about $\bar{\lambda}_L/4$ in these points, i.e.

$$H_T \approx \bar{\lambda}_L/4 \quad \text{for} \quad T = \hat{T}. \tag{4.369}$$

For the reflection coefficients these points indicate the midperiods of the decaying curves. For longer periods, the transmission coefficients b reach a limit which is different from 1. This could be surprising if we assumed that the eigenfunctions right and left of the discontinuity were to become identical for long periods with the concequence that the layer with its discontinuity would not be "noticed". However, in reality this is not the case. Already KNOPOFF et al. (1970) have pointed out that the eigenfunctions of Love waves contrary to Rayleigh waves are strong functions of the layer parameters for all periods including the infinite period. The surface transmission coefficients B according to Fig. 73 also have characteristic minima which are shifted, however, to longer periods in comparison to those of b. Obviously, the quantity K introduced by SICHARULIDZE (1978) is identical with b^2,

$$K = T_{BA} \cdot T_{BA} = b^2, \tag{4.370}$$

and $1 - K$ is plotted versus period in Fig. 75.

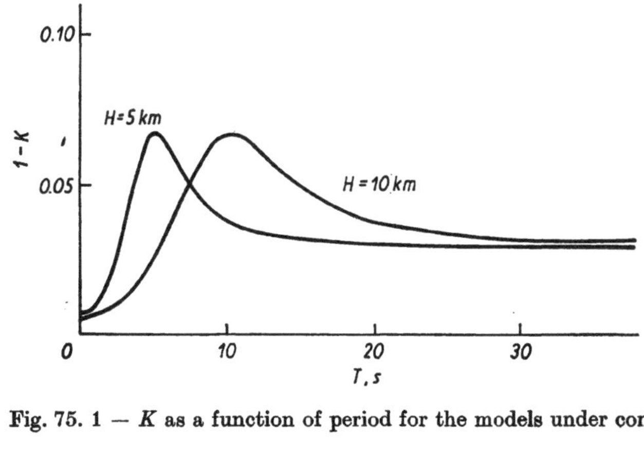

Fig. 75. $1 - K$ as a function of period for the models under consideration

This quantity which in principle can be extracted from the data has distinct maxima at the characteristic periods $\hat{T}$ which have to be connected somehow with the depth of the discontinuity. Such a connection is given for example by the rule of thumb (4.369) which is approximately valid for the considered models and parameter combinations only. But there is no common validity, even for the same model; but when using Rayleigh waves there will certainly be another relation. This can be also realized from the curves given by SICHARULIDZE (1978) which obviously refer to Rayleigh waves. While we find a qualitatively comparable picture in the intermediate period range, there are significant differences for the short period range: the distinct maxima according to Fig. 75 were not observed by SICHARULIDZE (1978). This fact could have various reasons. In each case, if we rely on the comparison with the exact solution according to Wiener-Hopf, we are sure that the curves given in Fig. 75 closely approach the exact ones, at the very least for periods greater than the limit period of the first higher mode. Numerous further model calculations for both Rayleigh and Love waves are necessary to make empirically well-founded statements about the depth of discontinuities using the method described here.

4.5.4. *Seismological applications of the theoretical conceptions concerning the reflection and transmission of surface waves*

There are comparatively few papers dealing with the analysis of seismic records with regard to vertical discontinuities and lateral inhomogeneities. Nevertheless, it was possible to confirm several theoretical statements experimentally. But we do not want to conceal that in some cases the occurring phenomena can be interpreted differently — a common situation in geophysics.

There are three basic phenomena we have to expect above all in analyzing the data:

a) Change of the phase and group velocity at the discontinuity as a result of the transition from one wave-guide into another one;
b) Occurrence of reflected wave groups and amplitude change of the transmitted waves;
c) Mode conversion.

All these phenomena are frequency dependent and precisely this fact is very important. As far as a) is concerned several authors assume that the phase velocity measured along a path with discontinuity is approximately equal to the mean of the phase velocities of the two structures joined together at the discontinuity. This statement was confirmed quite accurately by DRAKE and BOLT (1980) using a finite element model and interpreting seismograms of the Love wave fundamental mode which has crossed a continental margin. On the other hand, SCHLUE (1979) has likewise shown, using the method of finite elements, that phase velocity measurements carried out too close to a lateral interface between two geological zones may lead to completely wrong results.

Reflected wave groups have already been discussed in detail in section 4.5.2. At this point, let us refer to an existing discrepancy between data and theory. Generally, the reflection coefficients calculated with various approximation methods are essentially less than the observed ones for specific situations in nature. For seismic model experiments, however, there is more or less an agreement between theory and practice concerning the order of magnitude of the reflection coefficients [see. e.g. Rykunov and Fam van Tkhukh (1972), Malischewsky (1976) and Its and Yanovskaya (1979a, 1982). But we have to take into account that model experiments are carried out with high velocity contrasts to obtain accurately measurable reflections. Such high contrasts seldom are found in reality. Certainly, the reflection coefficient is increased by applying the method of Its and Yanovskaya (1982) on the basis of a refined Green's function; however, not nearly enough. The situation changes completely in the range of total reflection for oblique incidence. Here, large reflection coefficients are predicted theoretically too. Obviously, the model of a simple vertical discontinuity is not adequate for such situations in reality where large reflection coefficients are observed for approximately vertical incidence too. In the author's opinion, sloping and curved discontinuities, complicated horizontally layered structures with abruptly ending low-velocity channels as well as vertically layered structures with layer thicknesses which are a certain multiple of the wave length of the surface waves should be considered in solving this problem.

Concerning the amplitude variations of surface waves in crossing a discontinuity, a comparison of seismological data with theoretical calculations was carried out for the first time by McGarr (1969a). He used the records of Rayleigh waves from an ocean-bottom station near the San Andreas Fault and from some continental stations. McGarr's interpretation indicated that a large contrast in elastic constants and density at the continental margin is necessary to satisfy the data. Without a low-velocity zone, this contrast would be considerably greater. Gregersen and Alsop (1976) interpreted records of Love waves for the same geological configuration and carried out theoretical calculations. In this way it was confirmed that the existence of a low-velocity channel in the oceanic mantle is necessary to explain the observed trend of the transmission coefficient as a function of period. For the continental fundamental Love mode the main part of the energy is concentrated in the vicinity of the surface while most of the energy of the oceanic fundamental Love mode is transported within the low-velocity channel. This effect increases with decreasing period in the range from 100 s up to 20 s involving a decreasing transmission coefficient with increasing period for the direction ocean → continent. Such behaviour is very important, for example, in studying microseisms. By using a finite element model, Drake and Bolt (1980) pointed out that oceanic microseisms (fundamental-mode Love waves) can be reduced by approximately 4 on the continent. Irikura and Kawanaka (1980) succeeded in explaining spectral peculiarities of microseisms in Japan by applying similar theorectical conceptions.

Losses of Energy at the ocean-continent boundary have already been considered up to a certain degree by Mitchell et al. (1976) in studying the attenuation of fundamental-mode Love waves. However, these authors assumed constant transmission coefficients, and that is certainly not correct.

Niazy and Kazi (1980, 1981) referred to the fact that when the impedance contrast (do not confuse this "seismic impedance" with our impedance parameter ε!) across the lateral discontinuity becomes large, the amplitude of the transmitted wave from the hard medium into the soft one can be amplified by more than an order of magnitude at certain frequencies for geologically reasonable structures. This agrees with a well-known fact in earthquake engineering which is the amplification of earthquake ground motion by soft sedimentary basins.

One crucial point in interpreting seismograms of surface waves is the possible mode conversion at an interface. This fact was impressively demonstrated by Drake (1980) using a finite element model. For the chosen model, at periods shorter than 2 s, practically all the energy of the incident fundamental Love and Rayleigh modes is converted into higher modes, especially into the first higher one. The calculations of Gregersen and Alsop (1976) indicated that the mode conversion at the continental margin is important for greater periods in the range 20—25 s too, and it increases with decreasing period. As a consequence, the surface wave seismogram is distorted: the energy in the short-period part will be spread out in time, in such a way that the long periods become more and more prominent. The seismograms of the ocean-bottom station agree with this conception. Another application of mode conversion was presented by Kennett and Mykkeltveit (1984) in analyzing L_g-waves crossing the Central Graben of the North Sea. L_g-wave observations at NORSAR from a sequence of explosions show that the phase is almost extinguished by passing through this structure. To explain this effect the L_g propagation was modelled numerically by applying the theory of coupled modes (see section 4.5.3.2.6.). For this purpose, the reflection and transmission matrices were calculated for a 1 Hz wavetrain with 18 modes, and strong interactions between the modes in reflection as well as transmission were found. It is noticeable that reflection into the same mode number is often less efficient than into neighbouring modes. In transmission, the sedimentary basin model leaves more than 50 per cent of the energy transmitted in the original mode except for mode 1. Coupling between modes is not confined to nearest neighbours and can often extend over five or six other modes. The character of the modal train transmitted through the model is profoundly different from the incident wave field and the energy in L_g will be much reduced compared with paths which do not cross the feature. The overall effect is to smear out the S-wave energy over a wide range of group velocities and to transfer energy from the most energetic components to destroy the normal interference pattern which give rise to the characteristic L_g arrivals. In most circumstances where the L_g phase is not observed, crustal barriers to propagation are postulated but it is often difficult to be precise about their location. KENNETT and MYKKELTVEIT (1984) used a sequence of shots crossing the structural feature and so it was possible to identify the cause of the poor transmission of L_g-waves with observations at a distant receiver. GREGERSEN (1984) also reported about general observations of L_g propagation in the North Sea region. He succeeded in building up an areal picture of the pattern of crustal heterogeneity from the character of the L_g train. The frequency-dependent propagation of the L_g-waves was likewise explainable in terms of surface wave propagation across structural boundaries as well as in terms of anelastic attenuation.

For oblique incidence, the possible mode conversion between Rayleigh and Love waves was proved theoretically by MALISCHEWSKY (1976) and confirmed by GREGERSEN (1978) by carrying out further model calculations. This conversion is small at the continental margin but nevertheless significant around 20-s period. KENNETT and MYKKELTVEIT (1984) have observed that an explosive source producing purely P-waves builds up significant transverse components on the seismograms with travel time. This result is consistent with the theoretical prediction of progressive net transfer of energy from Rayleigh to Love modes as the wave train propagates through a three-dimensionally heterogeneous medium.

LEVSHIN (1984) discussed thoroughly the errors caused by neglecting lateral inhomogeneities between source and receiver in determining such important quantities as the attenuation Q or the seismic moment from surface wave data.

Recently, CHAMUEL (1984) pointed out that the presence of unexplained large attenuation of low-frequency components and inverted dispersion of Rayleigh waves crossing the ocean-continent margin, the Tibetan plateau, the Iranian plateau, and in general, steplike changes in elevation are examples of complex unsolved problems on Rayleigh wave propagation across vertical seismic boundaries. Using two- and three-dimensional ultrasonic models, these phenomena were explained. Laboratory seismic ultrasonic modelling results on Tibet were presented matching actual field data. Furthermore, physical iasights into the cause of the inverted dispersion and attenuation were obtained from the experimental results. Contrary to the current belief of the presence of a deep low velocity layer (possibly partially molten) at 30 km below the surface of the Tibetan Plateau, the new reported findings support Chamuel's concept of a low attenuation solid crustal structure with monotonically increasing velocity with depth. In each case, implications of thase findings include the need for new models of crustal structure in such regions taking into account the effects of topography.

5. The Wiener-Hopf technique

5.1. General remarks

A number of problems of mathematical physics can be solved using a method which is called "method of factorization" or "Wiener-Hopf technique". This method which was applied for the first time in solving integral equations in a paper of WIENER and HOPF (1931), allows the avoidance of analytical difficulties which occur in solving certain types of boundary value problems by integral transformations or by the transformation into integral equations. Above all, these problems refer to the diffraction of waves and to the theory of wave-guides, whose geometry is characterized by junctions of half-infinite subregions belonging to certain separable co-ordinate systems. By applying the Green's function technique to the Wiener-Hopf geometry an inhomogeneous integral equation, defined over a half-infinite range, of the general type

$$\nu f(x) + \int_0^\infty K(x-\eta)\, f(\eta)\, \mathrm{d}\eta = g(x) \quad \text{for} \quad 0 \leqq x < \infty \tag{5.1}$$

is obtained, where the kernel $K(x)$ is defined in the interval $-\infty < x < \infty$. The function $g(x)$ is defined for $0 \leqq x < \infty$, but $g(x)$ is undefined for $-\infty < x < 0$. The function $f(x)$ is identically zero for $-\infty < x < 0$, and $f(x)$ is to be determined in the range $0 \leqq x < \infty$. For several problems connected with electromagnetic wave-guides, $\nu \equiv 0$ is valid. The solution of this equation by the Wiener-Hopf technique is characterized by the application of the Fourier transformation and integration in the complex plane by using the process of analytical continuation and Liouville's theorem. The fact that the integration in equation (5.1) is carried out over positive η only, and that the equation holds for positive x only, is expressed by the analytical properties of the Fourier transform of two auxiliary functions. There is already an extensive literature dealing with the fundaments of the Wiener-Hopf technique. That is why we thoroughly want to treat here various details in applying this method to problems which are interesting in the context of the book. On this occasion, we prefer Jone's formulation, i.e. by-passing (5.1), the equations which are relevant to the Wiener-Hopf technique are obtained directly by applying the Fourier transformation to the partial differential equation. In our representation, which tentatively approaches the solution of the impedance jump problem in open as well as closed wave-guides, we follow by and large the book of WEINSTEIN (1969). It should be taken into account that the notations used in this chapter partly overlap with those in the other chapters. The definitions given previously remain valid in those cases where a quantity is not explained in the following.

5.2. Application of the Wiener-Hopf technique to an impedance jump in an open wave-guide

5.2.1. *Preliminary remarks and derivation of two functional equations*

For the homogeneous half-space, the fundamental equation of motion used by us is

$$\Delta u + k_\beta^2 u = 0, \qquad u = u(x_1, x_3), \tag{5.2}$$

or by using a more complicated operator $\tilde{\Delta}$ for the layered half-space (i.e. which is inhomogeneous in x_3-direction):

$$\tilde{\Delta} u = 0. \tag{5.3}$$

First let us assume that this half-space is extended in x_1-direction from $-\infty$ up to $+\infty$, and the fundamental solutions which are consistent with this geometry can be written for example in the following way

$$u = A\, e^{i(\pm k x_1 \pm \gamma x_3)} \quad \text{with} \quad \gamma = \sqrt{k_\beta^2 - k^2}, \tag{5.4}$$

where A is a constant not yet determined. In the presence of boundary conditions for $x_3 = 0$ (and $x_3 = \infty$), the spectrum of k-values is restricted by a secular equation

$$\Delta_s(k) = 0, \tag{5.5}$$

whose solutions are a sequence ${}^{\alpha}k$ $(\alpha = 0, 1, 2, \ldots)$ corresponding to the discrete normal modes plus the radiation modes. At the moment, the latter are not considered; later on they appear as an integral over a branch line in the complex plane. For analytic convenience we will assume that the medium is slightly lossy, that means that k_β has a small imaginary part. The purely lossless medium is considered as the limiting case $\operatorname{Im} k_\beta = 0$. In this way, we obtain a strip of regularity in applying the Wiener-Hopf technique. Later on, the Fourier representation of the solution requires an integration of the double-valued quantity γ [see (5.4)] in the complex plane. To specify γ unequivocally, a careful analysis of the branch cuts is necessary, and we will do it following the representation of MITTRA and LEE (1971). The specification of γ is obtained by invoking the radiation condition (compare with section 3.3.6.)

$$\operatorname{Re} \gamma \geqq 0 \tag{5.6}$$

and

$$\operatorname{Im} \gamma \geqq 0, \tag{5.7}$$

which specifies either an outgoing or evanescent wave in infinity. The function $\gamma = (k_\beta^2 - k^2)^{1/2}$ has branch points at $k = \pm k_\beta$ and we will lead the branch cuts from these points up to $\pm i\infty$. To specify γ unequivocally, we may view the complex k-plane ($k \overset{!}{=} k_r + ik_i$) as a two-sheeted surface with the sheets connected along the branch cuts. In each sheet γ is a single-valued analytic function of k. If we vary k continuously, the value of γ becomes discontinuous only if a branch cut is crossed, in which case the sign of γ is changed across the cut. Now we will discuss a particular branch cut choice based on the restrictions on γ expressed in (5.6) and (5.7). Let us assume that $\operatorname{Im} \gamma > 0$ in one sheet, say, the top one, while $\operatorname{Im} \gamma < 0$ in the bottom sheet. This definition implies that the two sheets are joined together by the curve $\operatorname{Im} \gamma = 0$, which therefore locates the desired branch cut. It is advisable to start from γ^2,

$$\gamma^2 = (k_\beta^{(r)2} - k_\beta^{(i)2}) - (k_r^2 - k_i^2) + 2i(k_\beta^{(r)}k_\beta^{(i)} - k_r k_i) \tag{5.8}$$

$$\text{with} \quad k_\beta = k_\beta^{(r)} + ik_\beta^{(i)},$$

and to divide the complex k-plane into regions bounded by curves $\operatorname{Re} \gamma^2 = 0$ or $\operatorname{Im} \gamma^2 = 0$ (see Fig. 76).

To ensure that $\operatorname{Im} \gamma > 0$ in the entire top sheet the inequality $|\arg \gamma^2| < 2\pi$ should hold there. This in turn dictates the choice of the cut to be

$$\arg \gamma^2 = 2\pi \triangleq \operatorname{Re} \gamma^2 > 0 \quad \& \quad \operatorname{Im} \gamma^2 = 0 \tag{5.9}$$

and thus the location of the branch cut is unequivocally determined (Fig. 77).

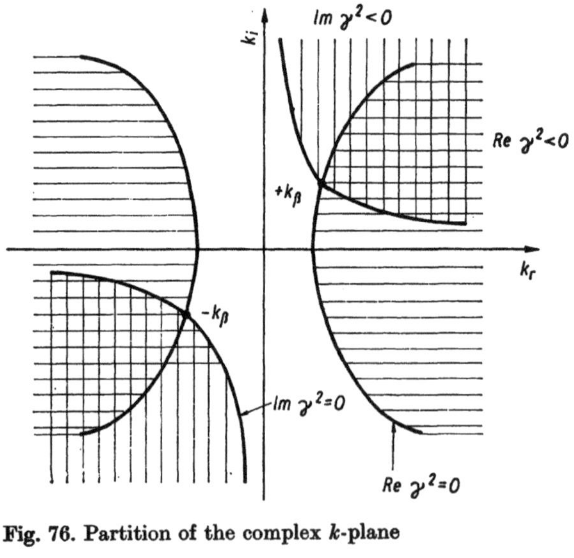

Fig. 76. Partition of the complex k-plane

Table 9 illustrates once again the different signs of Re γ and Im γ on both sheets.

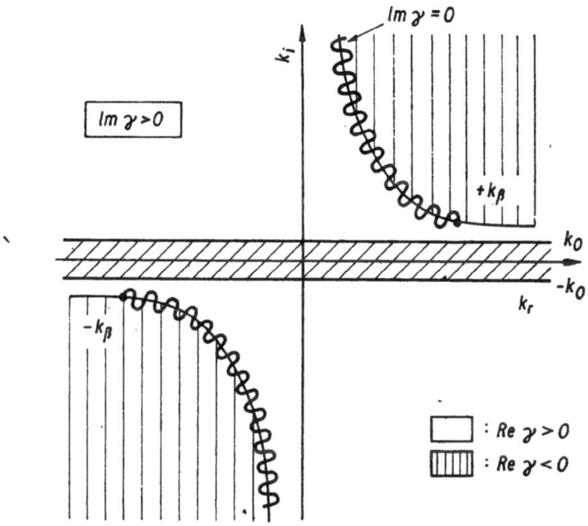

Fig. 77. Location of the branch cuts

If the integration path is interpreted as the one along the real axis in the top sheet, the radiation condition (5.6) and (5.7) is satisfied for every k on the path, and the integral representation is thus completely specified. It is important to realize that once the integral path is selected, it can then be deformed, according to the theory of complex variables, in an arbitrary manner on the two-sheeted k-plane.

Table 9. The different signs on both sheets

	Im γ	Re γ	
Top sheet	+	▥ (hatched)	−
	+	☐ (blank)	+
Bottom sheet	−	▥ (hatched)	+
	−	☐ (blank)	−

The equation of motion (5.2) together with the boundary conditions

$$\mu \frac{\partial u}{\partial x_3} + \varepsilon_1 u = 0 \quad \text{for} \quad x_1 < 0 \quad \text{and} \quad x_3 = 0, \tag{5.10}$$

$$\mu \frac{\partial u}{\partial x_3} + \varepsilon_2 u = 0 \quad \text{for} \quad x_1 > 0 \quad \text{and} \quad x_3 = 0, \tag{5.11}$$

defines the impedance jump problem, whose solution is sought using the statement

$$u = u_0 + \int\limits_{\mathscr{C}} e^{i(kx_1 + \gamma x_3)} \frac{F(k)}{\gamma - i\dfrac{\varepsilon_2}{\mu}} \, dk, \tag{5.12}$$

where u_0 is an incident surface wave

$$u_0 = C\, e^{ik_1 x_1 - \frac{\varepsilon_1}{\mu} x_3}, \tag{5.13}$$

$F(k)$ is a function to be determined, and $\mathscr{C}$ is a convenient integration path. In this way the problem is completely determined. In contrast to the mode-matching procedure, additional continuity conditions are not required. They are fulfilled automatically by the form of the statement. Obviously, the statement (5.12) satisfies the equation of motion (5.2). The incorporation of the boundary conditions (5.10) and (5.11) yields the two desired functional equations which will be the starting-point for the application of the Wiener-Hopf technique:

$$\text{I} \quad \int\limits_{\mathscr{C}} e^{ikx_1} \Psi(k)\, F(k)\, dk = 0 \quad \text{for} \quad x_1 < 0 \tag{5.14}$$

and

$$\text{II} \quad \int\limits_{\mathscr{C}} e^{ikx_1} F(k)\, dk = D\, e^{ik_1 x_1} \quad \text{for} \quad x_1 > 0, \tag{5.15}$$

where $\Psi(k)$ is an auxiliary function, defined by

$$\Psi(k) = \frac{\gamma - i\dfrac{\varepsilon_1}{\mu}}{\gamma - i\dfrac{\varepsilon_2}{\mu}}, \tag{5.16}$$

and D is a constant,

$$D = \frac{iC}{\mu} \Delta\varepsilon. \tag{5.17}$$

5.2.2. *Solution of the functional equations I and II*

The solution of the system I + II is easy, if a "*factorization*" of the function $\Psi(k)$ is known,

$$\Psi(k) = \Psi_+(k)\,\Psi_-(k), \tag{5.18}$$

where the function $\Psi_+(k)$ is holomorphic and has no zeros in the upper half-plane $\mathrm{Im}\,k \geqq 0$ and tends to one for $|k| \to \infty$, and $\Psi_-(k)$ has the same properties in the lower half-plane $\mathrm{Im}\,k \leqq 0$. Choosing $F(k)$ as

$$F(k) = \frac{D}{2\pi\mathrm{i}(k - k_1)}\,\frac{\Psi_-(-k_1)}{\Psi_+(k)}, \tag{5.19}$$

the functional equations I and II are satisfied. To prove this, we consider the following identity

$$\oint \frac{\mathrm{e}^{\mathrm{i}kx_1}\,\mathrm{d}k}{(k - k_1)\,\Psi_+(k)} = \int\limits_{\mathscr{C}} \cdots + \int\limits_{\mathscr{C}'} \cdots = \int\limits_{\mathscr{C}} \frac{\mathrm{e}^{\mathrm{i}kx_1}\,\mathrm{d}k}{(k - k_1)\,\Psi_+(k)} \tag{5.20}$$

with an integration path shown in Fig. 78.

Here, $\mathscr{C}$ corresponds to an integration along the real axis from $-\infty$ up to $+\infty$ and $\mathscr{C}'$ is a semi-circle with an infinite radius. According to Jordan's lemma the integral over $\mathscr{C}'$ vanishes and consequently (5.20) is valid. Since the integrand of the contour integral has a pole for $k = k_1$, it follows that

$$\oint \cdots = 2\pi\mathrm{i}\,\mathrm{Res}|_{k=k_1}. \tag{5.21}$$

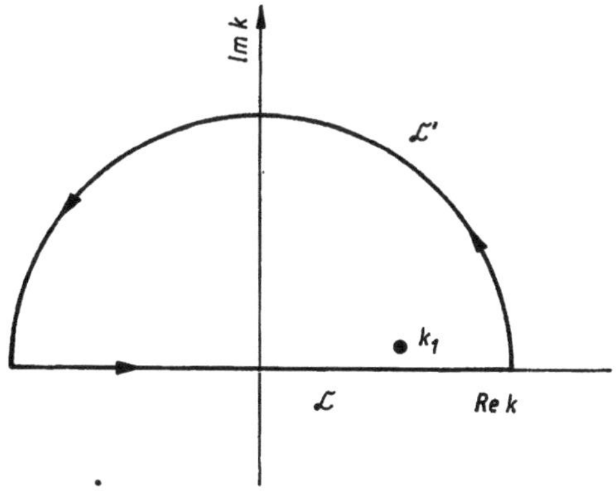

Fig. 78. The integration path for the identity (5.20)

By using the relation

$$\Psi_+(k) = \Psi_-(-k) \tag{5.22}$$

which has to be justified later on [compare with (5.34)], the residuum is obtained as

$$\text{Res}\left(\frac{e^{ikx_1}\,dk}{(k-k_1)\,\Psi_+(k)}\right)_{k=k_1} = \frac{e^{ik_1x_1}}{\Psi_-(-k_1)} \tag{5.23}$$

and consequently

$$\int_{\mathscr{C}} \frac{e^{ikx_1}}{(k-k_1)\,\Psi_+(k)} = 2\pi i\,\frac{e^{ik_1x_1}}{\Psi_-(-k_1)}. \tag{5.24}$$

The validity of equation I follows immediately by combining (5.15), (5.19), and (5.24). Furthermore, by substituting (5.18) and (5.19) into (5.14) it follows that

$$\int_{\mathscr{C}} e^{ikx_1}\Psi(k)\,F(k)\,dk = \frac{D}{2\pi i}\,\Psi_-(-k_1)\int_{\mathscr{C}} e^{ikx_1}\,\frac{\Psi_-(k)}{k-k_1}\,dk. \tag{5.25}$$

If the contour is closed with $\mathscr{C}'$ in the lower half-plane instead of the upper half-plane, then there is no pole inside the contour, the integral (5.25) vanishes identically, and consequently equation II is satisfied too, q.e.d.

5.2.3. *Factorization of the function $\Psi(k)$*

The necessity of splitting up a certain auxiliary function according to (5.18) is characteristic for the Wiener-Hopf technique and at the same time it is the most complicated part of the problem. Introducing the additional auxiliary functions

$$\psi_1(k) = 1 - i\,\frac{\varepsilon_1}{\mu\gamma}, \qquad \psi_2(k) = 1 - i\,\frac{\varepsilon_2}{\mu\gamma} \tag{5.26}$$

it follows that

$$\Psi(k) = \frac{\psi_1(k)}{\psi_2(k)} \tag{5.27}$$

and that

$$\Psi_+(k) = \frac{\psi_{1+}(k)}{\psi_{2+}(k)}, \qquad \Psi_-(k) = \frac{\psi_{1-}(k)}{\psi_{2-}(k)}, \tag{5.28}$$

so that the problem is reduced to the factorization of the simpler function ψ into ψ_+ and ψ_-, where ψ_+ is holomorphic and has no zeros in the upper half-plane $\operatorname{Im} k \geqq 0$ and ψ_- has the same properties in the lower half-plane $\operatorname{Im} k \leqq 0$. The functions ψ_{1+}, ψ_{2+}, ψ_{1-}, and ψ_{2-} are obtained by specifying ε. First, we proceed from the assumption that $\psi(k)$ is holomorphic and has no zeros in a strip of the complex plane $-k_0 \leqq \operatorname{Im} k \leqq k_0$, $0 < k_0 < \operatorname{Im} k_\beta$. To carry out the factorization, we consider a function $\chi(k)$ and split it into a sum such as:

$$\chi(k) = \log \psi(k) = \chi_+(k) + \chi_-(k), \tag{5.29}$$

so that $\chi_+(k)$ is holomorphic for $\operatorname{Im} k \geqq -k_0$ and $\chi_-(k)$ is holomorphic for $\operatorname{Im} k \leqq k_0$. The logarithmic function is specified in such a way that $\chi(k)$ goes to zero for $|k| \to \infty$ within the strip $-k_0 \leqq \operatorname{Im} k \leqq k_0$ (see Fig. 77). Such a choice of the branch of $\log \psi(k)$ is possible, because obviously $\psi(k)$ is an even function,

$$\psi(k) = \psi(-k), \tag{5.30}$$

tending to one for $\operatorname{Re} k \to \pm\infty$. Using a factorization theorem [see e.g. MITTRA and LEE (1971)] we obtain one of the possible decompositions of $\chi(k)$ as follows

$$\chi_+(l) = \frac{1}{2\pi \mathrm{i}} \int\limits_{-\mathrm{i}k_0-\infty}^{-\mathrm{i}k_0+\infty} \frac{\chi(k)\,\mathrm{d}k}{k-l}, \qquad \chi_-(l) = -\frac{1}{2\pi \mathrm{i}} \int\limits_{\mathrm{i}k_0-\infty}^{\mathrm{i}k_0+\infty} \frac{\chi(k)\,\mathrm{d}k}{k-l}. \tag{5.31}$$

According to (5.29) the auxiliary functions

$$\psi_+(k) = \mathrm{e}^{\chi_+(k)}, \qquad \psi_-(k) = \mathrm{e}^{\chi_-(k)} \tag{5.32}$$

satisfy the equation

$$\psi(k) = \psi_+(k)\, \psi_-(k) \tag{5.33}$$

and furthermore, as it can be shown using (5.31),

$$\psi_+(-k) = \psi_-(k), \tag{5.34}$$

and consequently (5.22) is justified, too. According to the factorization theorem ψ_+ is holomorphic and has no zeros in the half-plane $\operatorname{Im} k \geqq -k_0$ and tends to one for $|k| \to \pm\infty$, and ψ_- has the same properties in the half-plane $\operatorname{Im} k \leqq k_0$. These are exactly the properties required by us to construct the solution.

First it is advisable to proceed from the derivation of e.g. $\chi_-(l)$,

$$\frac{\mathrm{d}\chi_-(l)}{\mathrm{d}l} = -\frac{1}{2\pi \mathrm{i}} \int\limits_{\mathrm{i}k_0-\infty}^{\mathrm{i}k_0+\infty} \frac{\chi(k)\,\mathrm{d}k}{(k-l)^2}, \tag{5.35}$$

and to use the theorem from function theory

$$\oint \frac{f'(z)}{z - z_0}\, dz = \oint \frac{f(z)}{(z - z_0)^2}\, dz. \tag{5.36}$$

Certainly, this formula is valid for a closed integration path only. However, it can be shown that the integral over a sufficiently large semi-circle vanishes, so that the relation can be applied to our integration path. By using

$$\frac{d\chi(k)}{dk} = \frac{d}{dk} \log \psi(k) = \frac{\psi'}{\psi} \tag{5.37}$$

it follows from (5.35)

$$\frac{d\chi_-(l)}{dl} = -\frac{1}{2\pi i} \int\limits_{ik_0-\infty}^{ik_0+\infty} \frac{\chi(k)\, dk}{(k - l)^2} = -\frac{1}{2\pi i} \int\limits_{ik_0-\infty}^{ik_0+\infty} \frac{\chi'(k)}{k - l}\, dk$$

$$= -\frac{1}{2\pi i} \int\limits_{ik_0-\infty}^{ik_0+\infty} \frac{\psi'(k)}{\psi(k)} \frac{dk}{k - l}, \tag{5.38}$$

and by introducing

$$\psi(k) = 1 - i \frac{\varepsilon}{\mu \sqrt{k_\beta^2 - k^2}} \tag{5.39}$$

we obtain

$$\frac{d\chi_-(l)}{dl} = \frac{1}{2\pi i} \int\limits_{ik_0-\infty}^{ik_0+\infty} \frac{i\varepsilon}{\mu \left(\gamma - i \frac{\varepsilon}{\mu}\right)} \frac{k\, dk}{\gamma^2 (k - l)}, \qquad k_0 < \operatorname{Im} k_\beta. \tag{5.40}$$

Subsequently the integration path in the complex plane is deformed upwards, so that the cut $k_\beta \to i\infty$ is enclosed from both sides (see Fig. 79). Since l is situated below the integration path in accordance with the holomorphic domain of $\chi_-(l)$, only the pole of the integrand being the root of the equation

$$\gamma - i \frac{\varepsilon}{\mu} = 0, \tag{5.41}$$

is touched in deforming the contour. Defining

$$-i \frac{\varepsilon}{\mu k_\beta} \stackrel{!}{=} \cos \zeta \qquad \left(\operatorname{Re} \zeta = \frac{\pi}{2}\right) \tag{5.42}$$

the pole is situated at

$$k = k_\beta \sin \zeta \quad \text{and} \quad \gamma = -k_\beta \cos \zeta . \tag{5.43}$$

Furthermore, we have to take into account that the branch point $k = k_\beta$ itself is a pole too. Resulting from the deformation, the formula

$$\frac{d\chi_-}{dl} = \frac{1}{2\pi i} (2\pi i \sum \text{Res} + I) \tag{5.44}$$

is obtained, where I means the loop integral along the branch cut (see Fig. 79).

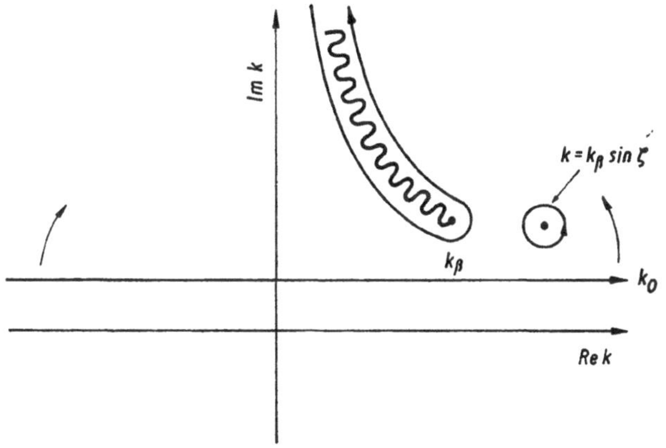

Fig. 79. Deformation of the original integration path

The residuum at $\gamma = -k_\beta \cos \zeta$ takes the value

$$\text{Res} \left. \frac{i\varepsilon k}{\mu \left(\gamma - i \frac{\varepsilon}{\mu}\right) \gamma^2 (k - l)} \right|_{\gamma = -k_\beta \cos\zeta} = \frac{1}{l - k_\beta \sin \zeta} \tag{5.45}$$

and for $\gamma = 0$

$$\text{Res} \left. \frac{1}{\gamma^2} \frac{i\varepsilon k}{\mu \left(\gamma - i \frac{\varepsilon}{\mu}\right) (k - l)} \right|_{\gamma = 0} = \frac{1}{2(k_\beta - l)}, \tag{5.46}$$

which follows immediately by applying the elementary rules of residuum calculus, so that the sought derivation of χ_- is for the present obtained as

$$\frac{d\chi_-}{dl} = \frac{1}{l - k_\beta \sin \zeta} + \frac{1}{2(k_\beta - l)} + \frac{I}{2\pi i}. \tag{5.47}$$

The evaluation of the loop integral I is essentially more complicated. First we proceed from the fact that according to Table 9 $\operatorname{Re} \gamma > 0$ holds on the lower border and $\operatorname{Re} \gamma < 0$ on the upper border and it follows that

$$I = \int_{i\infty}^{k_\beta} \frac{i\varepsilon k \, dk}{\mu\gamma^2 \left(\gamma - i\frac{\varepsilon}{\mu}\right)(k - l)} - \int_{k_\beta}^{i\infty} \frac{i\varepsilon k \, dk}{\mu\gamma^2 \left(\gamma + i\frac{\varepsilon}{\mu}\right)(k - l)}. \quad (5.48)$$

After elementary transformations we obtain

$$I = \frac{2\varepsilon}{i\mu} \int_{k_\beta}^{i\infty} \frac{k \, dk}{\gamma \left(\gamma^2 + \frac{\varepsilon^2}{\mu^2}\right)(k - l)}. \quad (5.49)$$

By applying the substitution

$$k = k_\beta \sin \tau, \quad \gamma = k_\beta \cos \tau, \quad l = k_\beta \sin \sigma; \quad dk = k_\beta \cos \tau \, d\tau \quad (5.50)$$

and using (5.42) the expression (5.49) can be transformed into

$$I = \frac{2\varepsilon i}{\mu k_\beta^2} \int_{\pi/2}^{i\infty} \frac{\sin \tau \, d\tau}{(\sin^2 \tau - \sin^2 \zeta)(\sin \tau - \sin \sigma)}. \quad (5.51)$$

The following integrals (see Appendix 7.1.)

$$I_1 = \int_{\pi/2}^{i\infty} \frac{d\tau}{\sin \tau - \sin \sigma} = i\,\frac{\pi/2 + \sigma}{\cos \sigma}, \quad (5.52)$$

$$I_2 = \int_{\pi/2}^{i\infty} \frac{d\tau}{\sin \tau - \sin \zeta} = -i\,\frac{\pi + \left(\frac{\pi}{2} - \zeta\right)}{\cos \zeta}, \quad (5.53)$$

$$I_3 = \int_{\pi/2}^{i\infty} \frac{d\tau}{\sin \tau + \sin \zeta} = i\,\frac{\pi/2 - \zeta}{\cos \zeta} \quad (5.54)$$

together with the trigonometric identity

$$\frac{2 \sin \tau}{(\sin^2 \tau - \sin^2 \zeta)(\sin \tau - \sin \sigma)}$$

$$= \frac{1}{\sin \sigma - \sin \zeta}\left(\frac{1}{\sin \tau - \sin \sigma} - \frac{1}{\sin \tau - \sin \zeta}\right)$$

$$+ \frac{1}{\sin \sigma + \sin \zeta}\left(\frac{1}{\sin \tau - \sin \sigma} - \frac{1}{\sin \tau + \sin \zeta}\right) \tag{5.55}$$

allow the writing of I in the manner

$$I = \frac{\varepsilon \mathrm{i}}{\mu k_\beta^2}\left[\frac{I_1 - I_2}{\sin \sigma - \sin \zeta} + \frac{I_1 - I_3}{\sin \sigma + \sin \zeta}\right] \tag{5.56}$$

or by taking into account (5.42) and (5.50)

$$I = -\frac{\pi \mathrm{i}}{l - k_\beta \sin \zeta} - \frac{\mathrm{i}}{k_\beta \cos \sigma}\left[\frac{(\pi/2 + \sigma) \cos \zeta + (\pi/2 - \zeta) \cos \sigma}{\sin \sigma - \sin \zeta}\right.$$

$$\left. + \frac{(\pi/2 + \sigma) \cos \zeta - (\pi/2 - \zeta) \cos \sigma}{\sin \sigma + \sin \zeta}\right]. \tag{5.57}$$

Applying an additional trigonometric identity

$$\sin^2 \sigma - \sin^2 \zeta = \sin(\sigma - \zeta) \sin(\sigma + \zeta) \tag{5.58}$$

the expression in the bracket can be simplified considerably as follows

$$I = -\frac{\pi \mathrm{i}}{l - k_\beta \sin \zeta} - \frac{\mathrm{i}}{k_\beta \cos \sigma}\left[\frac{\sigma + \zeta}{\sin(\sigma + \zeta)} - \frac{\pi + \sigma - \zeta}{\sin(\pi + \sigma - \zeta)}\right] \tag{5.59}$$

and substituting (5.59) into (5.47) finally the derivation of χ_- is obtained as

$$\frac{\mathrm{d}\chi_-(l)}{\mathrm{d}l} = \frac{1}{2} \frac{1}{l - k_\beta \sin \zeta} + \frac{1}{2(k_\beta - l)} - \frac{1}{2\pi k_\beta \cos \sigma}$$

$$\times \left[\frac{\sigma + \zeta}{\sin(\sigma + \zeta)} - \frac{\pi + \sigma - \zeta}{\sin(\pi + \sigma - \zeta)}\right]. \tag{5.60}$$

Because we need χ_- itself, this expression has to be integrated over l. To do this it is useful to change the integration variable for the third term in such a manner: $dl = k_\beta \cos\sigma\, d\sigma$, so that we obtain for the present

$$\chi_-(l) = \log \sqrt{\frac{\sin\zeta - \sin\sigma}{1-\sin\sigma}} - \frac{1}{2\pi}\int_0^\sigma \left[\frac{\sigma+\zeta}{\sin(\sigma+\zeta)} - \frac{\pi+\sigma-\zeta}{\sin(\pi+\sigma-\zeta)}\right] d\sigma \quad (5.61)$$

and after two further changes of variables

$$\chi_-(l) = \log \sqrt{\frac{\sin\zeta - \sin\sigma}{1-\sin\sigma}} - \frac{1}{2\pi}\left(\int_{\pi-\zeta+\sigma}^{\sigma+\zeta} \frac{t\,dt}{\sin t} + \int_\zeta^{\pi-\zeta} \frac{t\,dt}{\sin t}\right), \quad (5.62)$$

where the second integral is equal to a constant $\tilde{C}'$. Replacing σ by τ χ_- follows as a function of k in the manner

$$\chi_-(k) = \log \sqrt{\frac{\sin\zeta - \sin\tau}{1-\sin\tau}} - \frac{1}{2\pi}\int_{\tau-\zeta+\pi}^{\tau+\zeta} \frac{t\,dt}{\sin t} + \tilde{C}', \quad (5.63)$$

and by using (5.32) the ψ_--function sought by us is

$$\psi_-(k) = \tilde{C}\cdot \sqrt{\frac{\sin\zeta - \sin\tau}{1-\sin\tau}} \exp\left[-\frac{1}{2\pi}\int_{\tau-\zeta+\pi}^{\tau+\zeta} \frac{t\,dt}{\sin t}\right]. \quad (5.64)$$

The condition

$$\psi_-(-i\infty) = 1 \quad (5.65)$$

yields $\tilde{C} = 1$ and by taking into account (5.34) we obtain finally

$$\psi_+(k) = \sqrt{\frac{\sin\zeta + \sin\tau}{1-\sin\tau}} \exp\left[\frac{1}{2\pi}\int_{\tau+\zeta-\pi}^{\tau-\zeta} \frac{t\,dt}{\sin t}\right], \quad (5.66)$$

$$\psi_-(k) = \sqrt{\frac{\sin\zeta - \sin\tau}{1-\sin\tau}} \exp\left[-\frac{1}{2\pi}\int_{\tau-\zeta+\pi}^{\tau+\zeta} \frac{t\,dt}{\sin t}\right]. \quad (5.67)$$

These formulas can be verified directly, too. Firstly from $\varepsilon = 0$ (i.e. $\zeta = \pi/2$) it follows that $\psi_+ = \psi_- = 1$, as it should be, and secondly by using the identity

$$\frac{1}{2\pi}\left[\int_{\tau+\zeta-\pi}^{\tau-\zeta} \frac{t\,dt}{\sin t} + \int_{\tau+\zeta}^{\tau-\zeta+\pi} \frac{t\,dt}{\sin t}\right] = \frac{1}{2}\int_{\tau+\zeta}^{\tau-\zeta+\pi} \frac{dt}{\sin t} = \frac{1}{2}\log\frac{\cos\tau + \cos\zeta}{\cos\tau - \cos\zeta} \quad (5.68)$$

the product (5.33) can be evaluated in the manner

$$\psi_+\psi_- = 1 + \frac{\cos \zeta}{\cos \tau} = 1 - \mathrm{i}\frac{\varepsilon}{\mu\gamma} = \psi, \quad \text{q.e.d.} \tag{5.69}$$

With the proof of the relations (5.34), (5.65), and (5.69) we would have a complete foundation of the formulas (5.66) and (5.67). However, the factorization was presented here in great detail to demonstrate how it has to be carried out for similar functions.

5.2.4. *Determination of the reflection and transmission coefficients for an incident surface wave*

We proceed from a modified statement (5.12), where the integration path $\mathscr{C}$ is taken along the real axis with a little loop around the point k_1. First this path is deformed into the lower half-plane and in doing so the integral splits up into two residues (incident and reflected surface waves) and a loop integral (radiation modes or body waves). The integration path $\mathscr{C}_1$ of the latter one runs along the branch cut $-k_\beta \to -\mathrm{i}\infty$, but in contrast to Fig. 79 it is suitable to take it parallel to the imaginary axis (see Fig. 80).

As a result of the deformation of the integration path $\mathscr{C}$ we obtain the expression

$$u_1 = C\left[\mathrm{e}^{\mathrm{i}k_1x_1 - \frac{\varepsilon_1}{\mu}x_3} + R\mathrm{e}^{-\mathrm{i}k_1x_1 - \frac{\varepsilon_1}{\mu}x_3} + Q_1(x_1, x_3)\right], \tag{5.70}$$

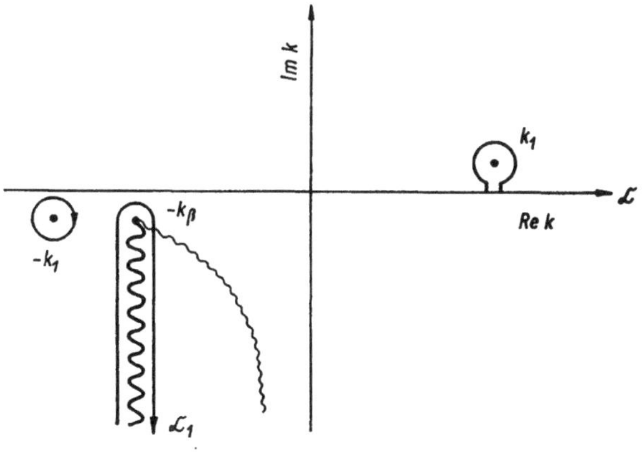

Fig. 80. Integration path for the determination of the reflection coefficient

in which Q_1 is given by

$$Q_1(x_1, x_3) = \frac{\Delta\varepsilon\Psi_-(-k_1)}{2\pi\mu} \int\limits_{\mathscr{C}_1} \frac{e^{i(kx_1+\gamma x_3)}\Psi_-(k)\,dk}{(k-k_1)\left(\gamma - i\dfrac{\varepsilon_1}{\mu}\right)}. \tag{5.71}$$

To evaluate the reflection coefficient R, we have to determine the residuum of the integrand of (5.12) at the point $k = -k_1$. Since according to (5.16) $\Psi(k)$ vanishes for $k = -k_1$, $\Psi_+(-k_1)$ has to vanish in the lower half-plane too, so that according to the rules of the residuum theory it follows that

$$\operatorname{Res} \left.\frac{D\Psi_-(-k_1)\,e^{i(kx_1+\gamma x_3)}}{\left(\gamma - i\dfrac{\varepsilon_2}{\mu}\right)(k-k_1)\,2\pi i\Psi_+(k)}\right|_{k=-k_1} = -\frac{D\Psi_-(-k_1)\,e^{-ik_1x_1-\frac{\varepsilon_1}{\mu}x_3}}{2\pi i\cdot 2k_1\cdot i\left(\dfrac{\varepsilon_1}{\mu}-\dfrac{\varepsilon_2}{\mu}\right)\Psi'_+(-k_1)}. \tag{5.72}$$

Furthermore,

$$\int\limits_{\mathscr{C}} \cdots = -2\pi i \operatorname{Res}|_{k=-k_1} + \text{loop integral} \tag{5.73}$$

holds and by using (5.17) and (5.70) the reflection coefficient can be written as

$$R = -\frac{\Psi_-(-k_1)}{2k_1\Psi'_+(-k_1)}. \tag{5.74}$$

Since according to (5.16) and (5.18)

$$\Psi'(k) = \frac{ik\Delta\varepsilon}{\mu\gamma\left(\gamma - i\dfrac{\varepsilon_2}{\mu}\right)^2} = \Psi'_+\Psi_- + \Psi_+\Psi'_- \tag{5.75}$$

is valid and as mentioned above $\Psi_+(-k_1) = 0$, we obtain

$$\Psi'_+(-k_1) = \frac{\Psi'(-k_1)}{\Psi_-(-k_1)} \tag{5.76}$$

and by taking into account (5.22) it follows from (5.74) that

$$R = -\frac{\varepsilon_1\Delta\varepsilon}{2k_1^2\mu^2}\Psi_+^2(k_1). \tag{5.77}$$

To obtain R explicitely it is first necessary by using (5.28) and (5.66) to evaluate $\Psi_+(k)$ in the manner:

$$\Psi_+(k) = \sqrt{\frac{\sin\zeta_1 + \sin\tau}{\sin\zeta_2 + \sin\tau}} \exp\left[\frac{1}{2\pi}\left(\int\limits_{\tau+\zeta_1-\pi}^{\tau-\zeta_1} \frac{t\,dt}{\sin t} + \int\limits_{\tau+\zeta_2-\pi}^{\tau-\zeta_2} \frac{t\,dt}{\sin t}\right)\right]. \tag{5.78}$$

For $k = k_1$ it follows from (5.42) and (5.50) that

$$\sin\tau = \sin\zeta_1 = \frac{k_1}{k_\beta}, \qquad \sin\zeta_2 = \frac{k_2}{k_\beta} \tag{5.79}$$

and furthermore

$$\zeta_1 = \frac{\pi}{2} + i\xi_1, \qquad \zeta_2 = \frac{\pi}{2} + i\xi_2 \tag{5.80}$$

with

$$\xi_1 = \operatorname{arcosh}\frac{k_1}{k_\beta} \qquad \text{and} \qquad \xi_2 = \operatorname{arcosh}\frac{k_2}{k_\beta}, \tag{5.81}$$

so that after substituting the integration variable $t \to is$ and after some transformations, the expression

$$\Psi_+(k_1) = \sqrt{\frac{2k_1}{k_1 + k_2}} \exp\left[\frac{i}{2\pi}\left(\int\limits_{\xi_1-\xi_2}^{\xi_1+\xi_2} \frac{s\,ds}{\sinh s} - \int\limits_{0}^{2\xi_1} \frac{s\,ds}{\sinh s}\right)\right] \tag{5.82}$$

is obtained. Consequently, by using (5.77) and (5.78) and by considering the identity

$$\frac{\varepsilon_1 - \varepsilon_2}{k_1 + k_2} = \mu^2 \frac{k_1 - k_2}{\varepsilon_1 + \varepsilon_2} \tag{5.83}$$

the reflection coefficient is given by

$$R = \frac{k_1 - k_2}{k_1} \cdot \frac{\varepsilon_1}{\varepsilon_1 + \varepsilon_2} \cdot \exp\left[\frac{i}{\pi}\left(\int\limits_{\xi_1-\xi_2}^{\xi_1+\xi_2} \frac{s\,ds}{\sinh s} - \int\limits_{0}^{2\xi_1} \frac{s\,ds}{\sinh s}\right)\right], \tag{5.84}$$

and its absolute value is represented by the very simple relation

$$|R| = \frac{|k_1 - k_2|}{k_1} \cdot \frac{\varepsilon_1}{\varepsilon_1 + \varepsilon_2}. \tag{5.85}$$

To evaluate the transmission coefficient, the integration path $\mathscr{C}$ is deformed into the upper half-plane and in doing so the integral splits up into a residuum (transmitted surface wave) and a loop integral (generated body waves). The integration path $\mathscr{C}_2$ of the latter one runs along the branch cut $k_\beta \to i\infty$, and it is again suitable to take it parallel to the imaginary axis (see Fig. 81).

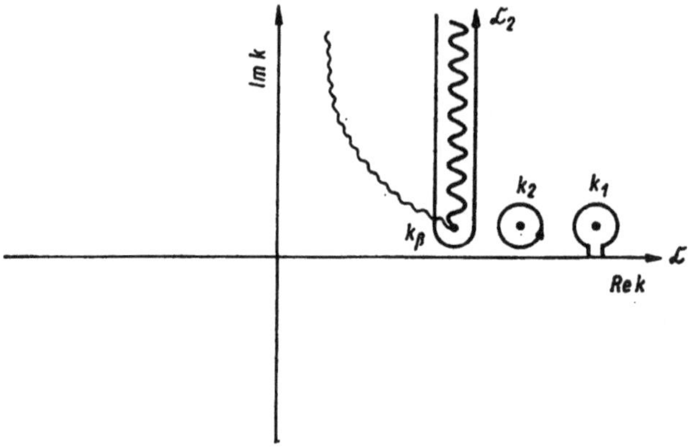

Fig. 81. Integration path for the determination of the transmission coefficient

As a result of the deformation of the integration path $\mathscr{C}$ we obtain analogously to (5.70) the expression

$$u_2 = C\left[T_r\, e^{ik_2x_1 - \frac{\varepsilon_2}{\mu}x_3} + Q_2(x_1, x_3)\right], \tag{5.86}$$

in which Q_2 is given by

$$Q_2(x_1, x_3) = \frac{\Delta\varepsilon\Psi_-(-k_1)}{2\pi\mu} \int\limits_{\mathscr{C}_2} \frac{e^{i(kx_1+\gamma x_3)}\, dk}{(k - k_1)\left(\gamma - i\frac{\varepsilon_2}{\mu}\right)\Psi_-(k)}. \tag{5.87}$$

To evaluate the transmission coefficient T_r we have to determine the residuum at the point $k = k_2$:

$$\operatorname{Res} \left.\frac{D\Psi_-(-k_1)\, e^{i(kx_1+\gamma x_3)}}{\left(\gamma - i\frac{\varepsilon_2}{\mu}\right)(k - k_1)\, 2\pi i\Psi_+(k)}\right|_{k=k_2}$$

$$= \left.\frac{D\Psi_-(-k_1)\, e^{i\left(k_2x_1 - \frac{\varepsilon_2}{\mu}x_3\right)}}{2\pi i(k_2 - k_1)\,\Psi_+(k_2)}\right| \operatorname{Res} \left.\frac{1}{\gamma - i\frac{\varepsilon_2}{\mu}}\right|_{k=k_2}$$

$$= -\frac{iC\Delta\varepsilon\varepsilon_2}{2\pi\mu^2k_2(k_2 - k_1)}\, \frac{\Psi_-(-k_1)}{\Psi_+(k_2)} \cdot e^{i\left(k_2x_1 - \frac{\varepsilon_2}{\mu}x_3\right)}. \tag{5.88}$$

Taking into account

$$\int_{\mathcal{C}} \cdots = 2\pi \mathrm{i} \, \mathrm{Res}|_{k=k_2} + \text{loop integral} \tag{5.89}$$

and the identities (5.22) and (5.83), the transmission coefficient follows from (5.86) and (5.88) as

$$T_r = \frac{k_1 + k_2}{\varepsilon_1 + \varepsilon_2} \cdot \frac{\varepsilon_2}{k_2} \cdot \frac{\Psi_+(k_1)}{\Psi_+(k_2)}. \tag{5.90}$$

If there is no surface wave in the second medium, obviously the field is represented there by Q_2 alone. For $k_1 = k_2$ it follows that $T_r = 1$ as it should be. The quantity $\Psi_+(k_2)$ is obtained similarly as $\Psi_+(k_1)$ from (5.78) by now setting $\tau = \zeta_2$,

$$\Psi_+(k_2) = \sqrt{\frac{k_1 + k_2}{2k_2}} \exp\left[\frac{\mathrm{i}}{2\pi}\left(\int\limits_0^{2\xi_2} \frac{s\,\mathrm{d}s}{\sinh s} - \int\limits_{\xi_2-\xi_1}^{\xi_2+\xi_1} \frac{s\,\mathrm{d}s}{\sinh s}\right)\right], \tag{5.91}$$

so that finally for T_r the expression

$$\begin{aligned} T_r = \frac{2\sqrt{k_1 k_2}}{k_2} \cdot \frac{\varepsilon_2}{\varepsilon_1 + \varepsilon_2} \cdot \exp\Bigg[\frac{\mathrm{i}}{2\pi}\Bigg(&\int\limits_{\xi_1-\xi_2}^{\xi_1+\xi_2} \frac{s\,\mathrm{d}s}{\sinh s} \\ &- \int\limits_0^{2\xi_1} \frac{s\,\mathrm{d}s}{\sinh s} - \int\limits_0^{2\xi_2} \frac{s\,\mathrm{d}s}{\sinh s} + \int\limits_{\xi_2-\xi_1}^{\xi_2+\xi_1} \frac{s\,\mathrm{d}s}{\sinh s}\Bigg)\Bigg] \end{aligned} \tag{5.92}$$

follows, whose absolute value has the simple form:

$$|T_r| = \frac{2\sqrt{k_1 k_2}}{k_2} \cdot \frac{\varepsilon_2}{\varepsilon_1 + \varepsilon_2}. \tag{5.93}$$

5.3. Application of the Wiener-Hopf technique to an impedance jump in a closed wave-guide

5.3.1. Preliminary remarks and derivation of the functional equations I and II

In the following, we will study the behaviour of a slow mode (i.e. a surface wave) at an impedance jump in a closed wave-guide type II [compare with (3.2)!]. To do this, it is necessary to modify the arguments of the last section. While the equation of motion (5.2) is valid as before, the boundary conditions are now given by

$$\mu \frac{\partial u}{\partial x_3} + \varepsilon_1 u = 0 \quad \text{for} \quad x_1 < 0 \quad \text{and} \quad x_3 = 0, \tag{5.94}$$

$$\mu \frac{\partial u}{\partial x_3} + \varepsilon_2 u = 0 \quad \text{for} \quad x_1 > 0 \quad \text{and} \quad x_3 = 0, \tag{5.95}$$

$$\frac{\partial u}{\partial x_3} = 0 \quad \text{for} \quad x_3 = H. \tag{5.96}$$

For the wave-guide without impedance jump, i.e. for the undisturbed wave-guide, the modified secular equation

$$\Delta_s^{\mathrm{II}}(k) = 0 \tag{5.97}$$

holds, whose solutions are unattenuated and attenuated fast modes and slow discrete ones as well. The fast modes are localized on the branch cuts according to Fig. 77. For the homogeneous wave-guide considered by us, there is one single slow mode only. Let this one be the incident wave u_0,

$$u_0 = C\, e^{i^0 k_1 x_1} \frac{\cosh\left[|^0\gamma_1|\,(x_3 - H)\right]}{\cosh |^0\gamma_1|\, H}, \tag{5.98}$$

where $^0\gamma_1$ is a solution of the secular equation (3.16) of medium 1 which applies to our case and $^0k_1$ follows from (3.17). The solution of the impedance jump problem in a closed wave-guide is sought again in the form of a statement,

$$u = u_0 + \int_{\mathscr{C}} e^{ikx_1} \frac{F(k) \cos\left[\gamma(x_3 - H)\right]}{\gamma \sin \gamma H + \frac{\varepsilon_2}{\mu} \cos \gamma H}\, dk, \tag{5.99}$$

which is certainly more complicated than for the open wave-guide. Nevertheless, by using the Wiener-Hopf technique a solution is successfully obtained in this case too. However, such a procedure would by no means be successful for all imaginable boundary conditions. The quantity $F(k)$ is again a function to be determined. The equation of motion and the boundary condition (5.96) are satisfied automatically by this statement, while the incorporation of the boundary conditions (5.94) and (5.95) yields, after some algebra, the well-known functional equations I and II [see (5.14) and (5.15)], in which the auxiliary function $\Psi(k)$ is now defined by

$$\Psi(k) = \frac{\gamma \sin \gamma H + \frac{\varepsilon_1}{\mu} \cos \gamma H}{\gamma \sin \gamma H + \frac{\varepsilon_2}{\mu} \cos \gamma H} \tag{5.100}$$

and the constant D is defined by

$$D = -\frac{\Delta\varepsilon}{\mu}\, C. \tag{5.101}$$

5.3.2. *Solution of the functional equations I and II and factorization of $\Psi(k)$*

If we know the factorization of $\Psi(k)$ according to (5.18), then $F(k)$ being a solution of the system I + II is obtained again by means of (5.19). To factorize $\Psi(k)$, we first introduce two auxiliary functions Ω_1 and Ω_2, defined by

$$\Omega_1 = 1 - \mathrm{i}\frac{\varepsilon_1}{\mu\gamma} - \left(1 + \mathrm{i}\frac{\varepsilon_1}{\mu\gamma}\right) \mathrm{e}^{2\mathrm{i}\gamma H}, \tag{5.102}$$

$$\Omega_2 = 1 - \mathrm{i}\frac{\varepsilon_2}{\mu\gamma} - \left(1 + \mathrm{i}\frac{\varepsilon_2}{\mu\gamma}\right) \mathrm{e}^{2\mathrm{i}\gamma H}, \tag{5.103}$$

so that $\Psi(k)$ can be written as

$$\Psi(k) = \frac{\Omega_1(k)}{\Omega_2(k)} \tag{5.104}$$

and the factorization of $\Psi(k)$ is reduced to the factorization of $\Omega(k)$ in the manner

$$\Psi_+(k) = \frac{\Omega_{1+}(k)}{\Omega_{2+}(k)}, \qquad \Psi_-(k) = \frac{\Omega_{1-}(k)}{\Omega_{2-}(k)}. \tag{5.105}$$

The factorization for the closed wave-guide can be described here in rough outlines only. For details we refer to the book of WEINSTEIN (1969). Actually, the procedure is analogous to the open wave-guide: the starting-point is again equation (5.38), in which Ω has to be introduced instead of ψ. A modified

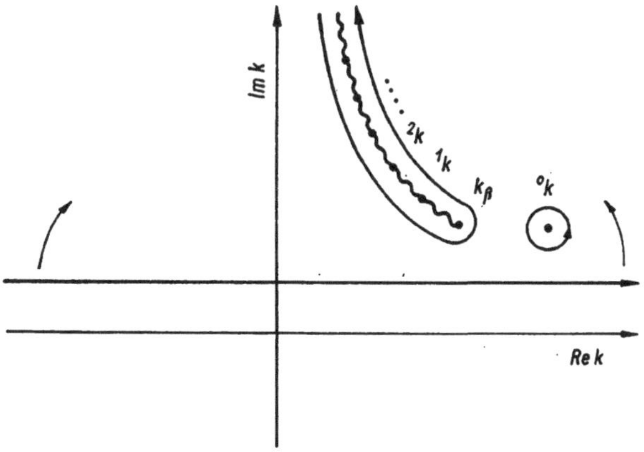

Fig. 82. Deformation of the integration path for the factorization of Ω

equation (5.40) follows, in which the integrand has an infinite number of poles ${}^{\nu}k$ ($\nu = 0, 1, 2, \ldots$) which are solutions of the secular equation (5.97) and which correspond to the modes of the closed wave-guide. The evaluation of the integral is carried out with the residuum calculus by deforming the original integration path into the upper half-plane. The integral is obtained as a sum of the residues and a loop integral which splits into parts along the segments between the poles at the upper and the lower border of the cut $k_\beta \to \mathrm{i}\infty$ (see Fig. 82). After a lot of calculation it follows that:

$$\Omega_+(k) = \hat{\Omega}\, \mathrm{e}^{\mathrm{i}qM_+(\tau)} \frac{1 + \dfrac{k}{{}^{0}k}}{\sqrt{1 + \dfrac{k}{k_\beta}}} \prod_{\nu=1}^{\infty} \left(1 + \frac{k}{{}^{\nu}k}\right) \mathrm{e}^{(\tau_\nu - \tau_{\nu-1})kH/\pi}, \tag{5.106}$$

$$\Omega_-(k) = \hat{\Omega}\, \mathrm{e}^{\mathrm{i}qM_-(\tau)} \frac{1 - \dfrac{k}{{}^{0}k}}{\sqrt{1 - \dfrac{k}{k_\beta}}} \prod_{\nu=1}^{\infty} \left(1 - \frac{k}{{}^{\nu}k}\right) \mathrm{e}^{(\tau_{\nu-1} - \tau_\nu)kH/\pi}, \tag{5.107}$$

where we have used the abbreviations

$$q = k_\beta H/\pi, \qquad \hat{\Omega} = \sqrt{\left(1 - \mathrm{i}\frac{\varepsilon}{\mu k_\beta}\right) \mathrm{e}^{-\mathrm{i}k_\beta H} - \left(1 + \mathrm{i}\frac{\varepsilon}{\mu k_\beta}\right) \mathrm{e}^{\mathrm{i}k_\beta H}} \tag{5.108}$$

and

$$k = k_\beta \sin \tau, \qquad M_+(\tau) = \left(\frac{\pi}{2} - \tau\right) \cos \tau + \sin \tau, \qquad M_-(\tau) = \left(\frac{\pi}{2} + \tau\right) \cos \tau - \sin \tau. \tag{5.109}$$

The quantities ${}^{\nu}k$ ($\nu = 0, 1, 2, \ldots$) are the roots of the equation

$$\Omega(k) = 0 \tag{5.110}$$

which corresponds to equation (5.97) and which is identical with (3.15), because

$$\tan x = \frac{\mathrm{e}^{2\mathrm{i}x} - 1}{\mathrm{i}(\mathrm{e}^{2\mathrm{i}x} + 1)} \tag{5.111}$$

holds. Furthermore, the quantities τ_ν are roots of equation (5.110) which have been transformed into the "τ-space", where τ_0 represents the slow mode and the relations

$$\tau_0 = \frac{\pi}{2} - \mathrm{i}\xi \tag{5.112}$$

with

$$\xi = \operatorname{arcosh} \frac{{}^0k}{k_\beta} \tag{5.113}$$

and

$${}^0k = \sqrt{k_\beta^2 + |{}^0\gamma|^2}, \qquad \tanh\left(|{}^0\gamma|\, H\right) = \frac{\varepsilon}{\mu\, |{}^0\gamma|} \tag{5.114}$$

are valid.

5.3.3. *Determination of the reflection coefficient for an incident slow mode*

First we substitute $F(k)$ according to (5.19) into the statement (5.99)

$$u = u_0 + \int_{\mathfrak{C}} \mathrm{e}^{\mathrm{i}kx_1} \cos\left[\gamma(x_3 - H)\right] \times \frac{D\Psi_-(-{}^0k_1)\, \mathrm{d}k}{2\pi\mathrm{i}(k - {}^0k_1)\, \Psi_+(k) \left[\gamma \sin \gamma H + \frac{\varepsilon_2}{\mu} \cos \gamma H\right]}, \tag{5.115}$$

or by using (5.105)

$$u = u_0 + \frac{D}{2\pi\mathrm{i}} \int_{\mathfrak{C}} \mathrm{e}^{\mathrm{i}kx_1} \frac{\cos\left[\gamma(x_3 - H)\right]}{(k - {}^0k_1)} \times \frac{\Omega_{1-}(-{}^0k_1)\, \Omega_{2+}(k)}{\Omega_{2-}(-{}^0k_1)\, \Omega_{1+}(k)} \frac{\mathrm{d}k}{\left[\gamma \sin \gamma H + \frac{\varepsilon_2}{\mu} \cos \gamma H\right]}. \tag{5.116}$$

Furthermore, by using the expressions for the Ω-function given under (5.106) and (5.107) and introducing the abbreviation p, defined by

$$p = p(k) = \frac{\prod\limits_{\nu=1}^{\infty} \left(1 + \frac{{}^0k_1}{{}^\nu k_1}\right) \mathrm{e}^{(\tau_\nu^{(1)} - \tau_{\nu-1}^{(1)})\, {}^0k_1 H/\pi} \prod\limits_{\nu=1}^{\infty} \left(1 + \frac{k}{{}^\nu k_2}\right) \mathrm{e}^{(\tau_\nu^{(2)} - \tau_{\nu-1}^{(2)}) kH/\pi}}{\prod\limits_{\nu=1}^{\infty} \left(1 + \frac{{}^0k_1}{{}^\nu k_2}\right) \mathrm{e}^{(\tau_\nu^{(2)} - \tau_{\nu-1}^{(2)})\, {}^0k_1 H/\pi} \prod\limits_{\nu=1}^{\infty} \left(1 + \frac{k}{{}^\nu k_1}\right) \mathrm{e}^{(\tau_\nu^{(1)} - \tau_{\nu-1}^{(1)}) kH/\pi}}, \tag{5.117}$$

we obtain

$$u = u_0 + \frac{D}{\pi \mathrm{i}} \int\limits_{\mathscr{C}} \mathrm{e}^{\mathrm{i}kx_1} \frac{\cos[\gamma(x_3 - H)]}{(k - {}^0k_1)}$$

$$\times \frac{p}{\left(1 + \frac{k}{{}^0k_1}\right)} \frac{{}^0k_2 + k}{({}^0k_2 + {}^0k_1)} \frac{\mathrm{d}k}{\left[\gamma \sin \gamma H + \frac{\varepsilon_2}{\mu} \cos \gamma H\right]}. \tag{5.118}$$

For the determination of the reflection coefficient, the integration path $\mathscr{C}$ is deformed into the lower half-plane, where no branch cut has to be taken into account. This is a result of the manner in which the integrand depends on the radical γ. There is an infinite number of poles only $k = -{}^0k_1, -{}^1k_1, -{}^2k_1, \ldots$ corresponding to the reflected modes which have to be considered in the form of residues. Taking into account

$$\left. \frac{D}{\gamma \sin \gamma H + \frac{\varepsilon_2}{\mu} \cos \gamma H} \right|_{k=-{}^\nu k_1} = \frac{C}{\cos {}^\nu\gamma_1 H} \tag{5.119}$$

we obtain the following result

$$u = C \left\{ \mathrm{e}^{\mathrm{i}{}^0k_1x_1} \frac{\cosh[|{}^0\gamma_1| (x_3 - H)]}{\cosh |{}^0\gamma_1| H} + \sum_{\nu=0}^{\infty} R_\nu \, \mathrm{e}^{-\mathrm{i}{}^\nu k_1 x_1} \frac{\cos[{}^\nu\gamma_1(x_3 - H)]}{\cos {}^\nu\gamma_1 H} \right\}, \tag{5.120}$$

in which the reflection coefficients R_ν are given by

$$R_\nu = 2 \operatorname{Res} \left. \frac{p({}^0k_2 + k)}{(k - {}^0k_1)\left(1 + \frac{k}{{}^0k_1}\right)({}^0k_2 + {}^0k_1)} \right|_{k=-{}^\nu k_1}, \quad \nu = 0, 1, 2, \ldots \tag{5.121}$$

Consequently, the reflection coefficient R_0 of the "slow mode in itself" is

$$R_0 = \frac{{}^0k_1 - {}^0k_2}{{}^0k_1 + {}^0k_2} \cdot p(-{}^0k_1). \tag{5.122}$$

The occurrence of infinite products at the reflection and transmission coefficients is a characteristic property of closed wave-guides [see e.g. TALANOV (1958)]. If only the slow mode can propagate in both wave-guides 1 and 2, i.e. ${}^\nu k_1$ and ${}^\nu k_2$ are purely imaginary for $\nu = 1, 2, \ldots$, the absolute value of the reflection coefficient according to (5.122) and (5.117) is simplified considerably:

$$|R_0| = \frac{|{}^0k_1 - {}^0k_2|}{{}^0k_1 + {}^0k_2}. \tag{5.123}$$

If the first two modes $\nu = 0,1$ can propagate we obtain the following expression from (5.122):

$$|R_0| = \frac{|{}^0k_1 - {}^0k_2|}{{}^0k_1 + {}^0k_2} \cdot \frac{{}^1k_1 + {}^0k_1}{|{}^1k_1 - {}^0k_1|} \cdot \frac{|{}^1k_2 - {}^0k_1|}{{}^1k_2 + {}^0k_1} \text{ and so on.} \tag{5.124}$$

The same problem for electromagnetic surface waves was solved by Miller and Talanov (1961) and the energy transmission coefficient which is analogous to (5.123) was found as

$$|T_0| = \frac{2\sqrt{{}^0k_1 {}^0k_2}}{{}^0k_1 + {}^0k_2}. \tag{5.125}$$

6. The three-dimensional problem

6.1. General remarks

A comprehensive treatment of the scattering of surface waves at discontinuities in 3D-structures is beyond the scope of this book. Here we want to refer briefly to some essential characteristics regarding the non-vertical (concerning the discontinuity) incidence of surface waves only. Obviously, this case is characterized by a greater complexity than the simple 2D-situation analogous to the behaviour of body waves. To cover oblique incidence it is necessary to include an exponential in the field vector (4.4) expressing the x_2-dependence. Subsequently, the boundary conditions at the discontinuity are fulfilled on the basis of Snell's law. To do this we have to apply the same approximation techniques as for the two-dimensional case. Rulf (1977b) was able to show that certain interesting physical effects, which occur at oblique incidence, may be discussed in qualitative terms without performing the whole analysis. All Rulf used in his analysis was a matching of x_2-components of the velocities. Whenever there are conditions for surface waves of different velocities on the two halves of the surface of an elastic half-space, the phenomena of refraction, total reflection, and non-radiation of one or both of the bulk waves occur. A very important point is the fact that alternative waves (Rayleigh → Love and vice versa) may arise, with corresponding additional critical angles. But contrary to the opinion of Rulf (1977b), these additional waves may also arise if the layer is approximated by surface boundary conditions as in section 2.1.2. Beyond that, the author was able to show that the neglecting of such additional waves leads to contradictions in eliminating the reflection and transmission coefficients according to section 4.4.3. Using a dual Wiener-Hopf formulation Chang and Kuester (1981) investigated a TEM surface wave mode obliquely incident onto the open end of a parallel plate wave-guide with a truncated upper plate. The reflection coefficient from the wave-guide opening was determined as a function of the incident angle and it was found that total reflection, i.e., a reflection coefficient with magnitude equal to unity, is possible for incident angles greater than the critical angle associated with the lowest-order surface wave mode of the grounded dielectric slab.

Furthermore, if a surface wave impinges on a medium with higher surface wave velocity, the occurrence of "surface-like" head waves is imaginable.

6.2. Some special cases

The magnitudes of the reflection and transmission coefficients of surface waves are complicated functions of the angle of incidence Θ as for electromagnetic waves according to Fresnel's formulas. First, let us demonstrate this using a simple impedance jump model, whose parameters are given in Fig. 83. The calculations are based on the functional F_2 of the LQBR-method and they were carried out neglecting alternative waves for both directions of propagation $1 \to 2$ and $2 \to 1$ respectively.

A critical angle Θ_{crit} occurs at 61,3° for the direction $1 \to 2$, but there is no such angle for the opposite direction. The course of the reflection coefficient is characterized by zero-crossings indicating phase jumps in these points.

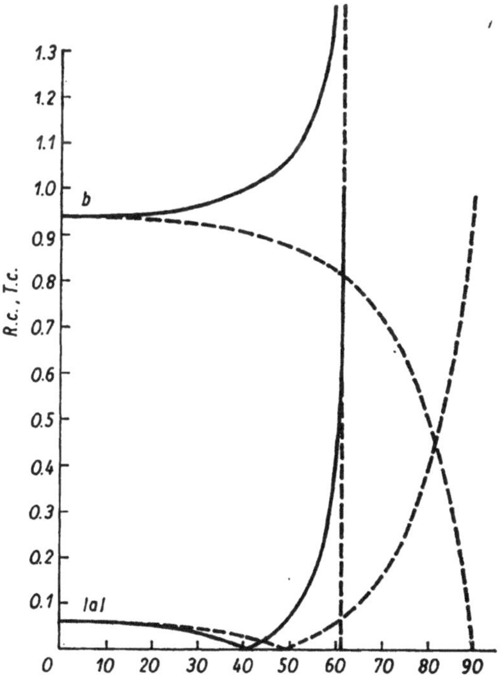

Fig. 83. Reflection and transmission coefficients for an impedance jump model as a function of the angle of incidence; full lines: $1 \to 2$, dashed lines: $2 \to 1$; Model parameters: $\omega = 1$, $\varrho = 3$, $\mu = 75$, $\varkappa_2 = \varkappa_3 = 1$;
$1 \to 2$: $\varkappa_1 = 0.5$, $\varepsilon = 10$
$2 \to 1$: $\varkappa_1 = 2$, $\varepsilon = 5$

Note that imaginary parts of the coefficients are again neglected in applying the LQBR-method. The transmission coefficients are energetic ones and that is why we cannot accept values greater than 1, as is the case in the range beyond the zero-crossing of the reflection coefficient for the direction $1 \to 2$. Similar theoretical results for Rayleigh waves and two half-spaces welded together are compared with experimental ones e.g. by ITS and YANOVSKAYA (1982). They pointed out that there is no agreement between theoretical and experimental values for the reflection coefficient in the vicinity of the critical angle. This is not surprising because it is experimentally very difficult to separate the "pure" reflected wave in this range.

Another point worth mentioning is the behaviour of the reflection coefficients for different material contrast at the discontinuity. Contrary to the opinion of GEGECHKORI and YANOVSKAYA (1979) the author [see MALISCHEWSKY and NEUNHÖFER (1983)] was able to show that it is not the velocity contrast which is responsible for the existence or non-existence of a zero-crossing, but the density contrast. In Fig. 84, we represent the results concerning Rayleigh waves and two half-spaces welded together for various density values in the second medium. Such behaviour is thus far interesting, because the density is a parameter which is not well determined from investigations concerning the dispersion of surface waves.

SCHLUE (1981) investigated the propagation of seismic surface waves in

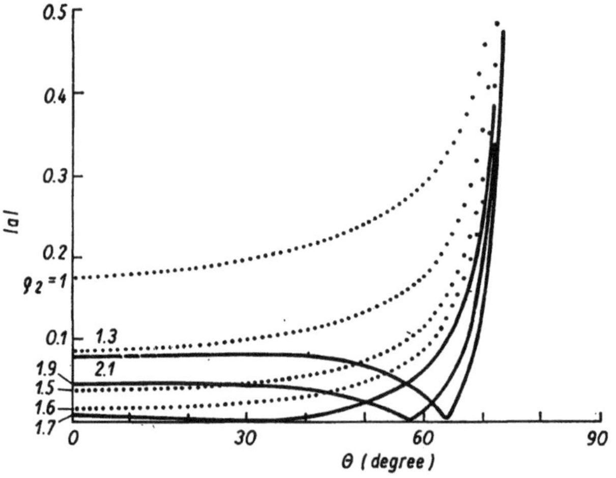

Fig. 84. Reflection coefficients for Rayleigh waves as a function of the angle of incidence and of the density contrast

Model 1: $\alpha_1 = 2.7$ km/s, $\beta_1 = 1.35$ km/s, $\varrho_1 = 1.7$ g/cm^3

Model 2: $\alpha_2 = 3.05$ km/s, $\beta_2 = 1.4$ km/s, $\varrho_2 = 1.0 \cdots 2.1$ g/cm^3

three-dimensional finite-element structures. The results from a model of a continent-ocean transition with angles of incidence 0° and 45° agree, in general, with those of GREGERSEN and ALSOP (1974) and GREGERSEN (1978), but disagree in some important details. So SCHLUE (1981) found no evidence of mode conversions between Rayleigh- and Love-wave modes, although such conversions are allowed by the finite-element formulation. This is not understandable, because these conversions should really exist.

Recently, BULLITT and TOKSÖZ (1985) reported about a three-dimensional ultrasonic modeling of Rayleigh wave scattering at topographic features. In accordance with our theoretical investigations they found that, in general, backscattered or reflected Rayleigh waves are small, compared with transmitted waves. However, a significant fraction of the Rayleigh wave energy is scattered into body waves. At oblique incidence angles, the transmission coefficients computed from spectral ratios vary strongly with incidence angle. Furthermore, for a vertical step, at each angle, the transmission coefficients vary greatly with frequency. Because of phase shifts, the wave train is distorted. According to BULLITT and TOKSÖZ (1985) some features of crustal Rayleigh waves commonly attributed to multipathing in the Earth may be due to scattering effects.

6.3. Some remarks concerning Rayleigh waves

Within the scope of this book we have concentrated upon two-dimensional models and Love-kind motion. But as mentioned above, the incorporation of angles of incidence other than zero (3D-problems) requires the treatment of Rayleigh waves built up by two perpendicular displacement vectors as well. Both the free surface and the impedance surface considered by us are capable of guiding Rayleigh waves (see section 2.1.2.), but a detailed discussion of these questions is beyond the scope of this book. Let us refer here to some peculiarities only. Following Kazi and Willmer [see KAZI and WILLMER (1981) and WILLMER and KAZI (1983)], which analyzed the eigenvalue problem associated with coupled seismic (P-SV) wave propagation in a uniform half-space, the two components of the displacement field satisfy a fourth order ordinary differential equation along with appropriate boundary conditions. The operator associated with the fourth order differential equation is the so-called Rayleigh wave operator, defined by

$$L_R = \frac{\mathrm{d}^4}{\mathrm{d}x_3^4} + (\gamma_\alpha^2 + \gamma_\beta^2)\frac{\mathrm{d}^2}{\mathrm{d}x_3^2} + \gamma_\alpha^2\gamma_\beta^2, \tag{6.1}$$

where γ_β is identical with γ from (3.4) and γ_α is defined by

$$\gamma_\alpha^2 = k_\alpha^2 - k^2, \qquad k_\alpha^2 = \frac{\omega^2}{\alpha^2} = \frac{\omega^2 \varrho}{2\mu + \lambda}. \tag{6.2}$$

With the displacement components in x_1 and x_3-directions U_1 and U_3 respectively, the eigenvalue problem is given by

$$L_R U_\tau = 0 \qquad (\tau = 1,3) \tag{6.3}$$

together with the two pairs of boundary conditions

$$B_1(U_1)\colon \frac{\mathrm{d}^2 U_1}{\mathrm{d}x_3^2} = \left[2k^2 \frac{(\gamma_\beta^2 - \gamma_\alpha^2)}{(\gamma_\beta^2 + k^2)} - \gamma_\beta^2\right] U_1 \quad \text{at} \quad x_3 = 0, \tag{6.4}$$

$$B_1(U_3)\colon \frac{\mathrm{d}^2 U_3}{\mathrm{d}x_3^2} = \left[2k^2 \frac{(\gamma_\alpha^2 - \gamma_\beta^2)}{(\gamma_\beta^2 + k^2)} - \gamma_\alpha^2\right] U_3 \quad \text{at} \quad x_3 = 0, \tag{6.5}$$

$$B_2(U_1)\colon \frac{\mathrm{d}^3 U_1}{\mathrm{d}x_3^3} = \left[2k^2 \frac{(\gamma_\beta^2 - \gamma_\alpha^2)}{(\gamma_\beta^2 + k^2)} + \gamma_\alpha^2 - 2\gamma_\beta^2)\right] \frac{\mathrm{d}U_1}{\mathrm{d}x_3} \quad \text{at} \quad x_3 = 0, \tag{6.6}$$

$$B_2(U_3)\colon \frac{\mathrm{d}^3 U_3}{\mathrm{d}x_3^3} = \left[\frac{k^2\gamma_\beta^2 + 2\gamma_\alpha^4 - \gamma_\beta^4}{\gamma_\beta^2 - 2\gamma_\alpha^2 - k^2}\right] \frac{\mathrm{d}U_3}{\mathrm{d}x_3} \quad \text{at} \quad x_3 = 0 \tag{6.7}$$

and a suitable condition at infinity also has to be imposed, namely that U_τ be square integrable. The spectral parameter k^2 occurs quadratically in the operator and linearly in the boundary conditions, whence we have a generalized eigenvalue problem which is not self-adjoint. ROKHLIN (1980) referred to this fact as well in investigating the diffraction of Lamb waves by a finite crack in an elastic layer. He pointed out that the spectrum of eigenwave numbers consists of a finite number of both real and pure imaginary wavenumbers, as well as an infinite set of complex wavenumbers. The complex nature of the Lamb-waves' spectrum is not the result of dissipation of energy in the layer's material, but of the mutual transformation of two different types of elastic waves (longitudinal and transverse) on the layer's surface. The non-propagating waves define the field near the inhomogeneity and, in particular, ensure the satisfaction of conditions at the edge of the crack. Regardless of some progress concerning Rayleigh and Lamb waves [see the citated papers of Kazi and Willmer] the final spectral representation of L_R has not been found up to now. In the author's opinion the eigenfunctions are complete for this problem too, but as pointed out by WILLMER and KAZI (1983) the completeness property will be difficult to prove in view of the inherent complexity of the non self-adjoint generalized eigenvalue problem under consideration.

7. Appendix

7.1. Calculation of the integrals I_1, I_2, I_3

The starting-point is the integral

$$\int \frac{d\tau}{\sin\tau - \sin\sigma} = \frac{1}{\cos\sigma} \log\left(-\tan\frac{\sigma}{2}\cdot\frac{\sin\frac{\tau-\sigma}{2}}{\cos\frac{\tau+\sigma}{2}}\right) + \bar{C}(\sigma), \tag{7.1}$$

which can be transformed into

$$\int \frac{d\tau}{\sin\tau - \sin\sigma} = \frac{1}{\cos\sigma} \log\frac{\sin\frac{\tau-\sigma}{2}}{\cos\frac{\tau+\sigma}{2}} \tag{7.2}$$

by choosing the constant $\bar{C}$ conveniently. Due to $k = k_\beta \sin\tau$ the integration along the lower border of the hyperbolic cut $k_\beta \to i\infty$ (5.49) corresponds to an

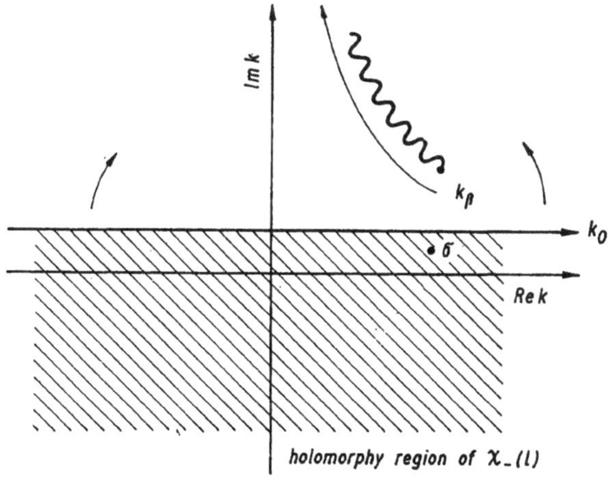

Fig. 85. The point σ and the holomorphy region of $\chi_-(l)$

integration in the τ-plane from $\frac{\pi}{2}$ up to $i\infty$ along a contour which has yet to be specified. First note, that the point σ is situated *below* the original integration path of (5.40) in conformity with the holomorphy region of $\chi_-(l)$ (see Fig. 85).

The transformation of the integration path into the τ-plane is illustrated in Fig. 86. We consider a closed contour which is formed by the hyperbolic cut transformed into the τ-plane (dashed line), a segment of a curve R_∞ in infinity, and a line which is parallel to the imaginary axis.

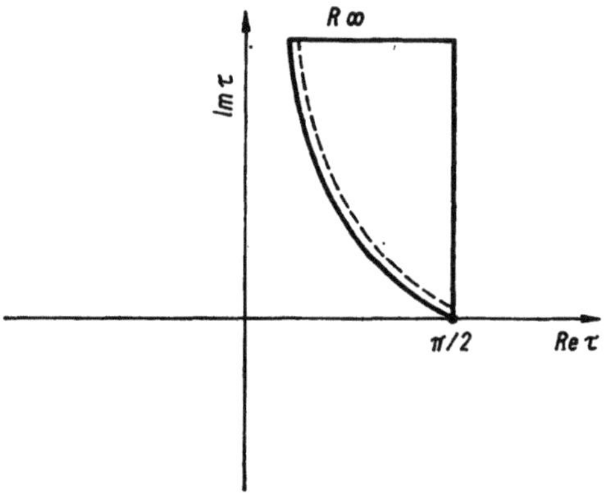

Fig. 86. The contour in the τ-plane

There are no singularities within the contour so that

$$\int_{i\infty}^{\pi/2} \cdots + \int_{\pi/2}^{i\infty+\pi/2} \cdots + \int_{R\infty} = 0 \tag{7.3}$$

holds. Since $\pi/2$ is "majorized" by $i\infty$ in the upper limit of the second integral of (7.3) and the integral over R_∞ vanishes, the integral along the dashed line can be replaced by the integral parallel to the imaginary axis. Consequently, using the identity

$$\frac{\sin\dfrac{\tau-\sigma}{2}}{\cos\dfrac{\tau+\sigma}{2}} = \frac{1-e^{i(\sigma-\tau)}}{i(e^{i\sigma}+e^{-i\tau})} \tag{7.4}$$

we obtain

$$I_1 = \int\limits_{\pi/2}^{i\infty} \frac{d\tau}{\sin\tau - \sin\sigma} = \frac{1}{\cos\sigma}\left[\log\frac{e^{i\tau} - e^{i\sigma}}{i(1 + e^{i\sigma}e^{i\tau})}\right]_{\pi/2}^{i\infty}$$

$$= \left[\log\frac{e^{-\infty} - e^{i\sigma}}{i(1 + e^{i\sigma}e^{-\infty})} - \log\frac{i - e^{i\sigma}}{i(1 + i\,e^{i\sigma})}\right]$$

$$= \frac{1}{\cos\sigma}\left[\log(i\,e^{i\sigma}) - \log 1\right] = \frac{1}{\cos\sigma}\left[\log(e^{i\pi/2}\,e^{i\sigma})\right]$$

$$= i\cdot\frac{\pi/2 + \sigma}{\cos\sigma} \qquad \text{q. e. d.} \tag{7.5}$$

The choice of the correct branch of the logarithmic function can be controlled by determining the integral for $\sigma = 0$. In this case, the integration path is taken along the real axis (from $\tau = \pi/2$ up to $\tau = 0$) and the imaginary axis (from $\tau = 0$ up to $\tau = i\infty$). Due to

$$\lim_{\delta\to 0}\left\{\int\limits_{\pi/2}^{\delta}\frac{d\tau}{\sin\tau} + \int\limits_{\delta}^{\infty}\frac{d\tau}{\sinh\tau}\right\} = \lim_{\delta\to 0}\log\frac{\tan(\delta/2)}{\tanh(\delta/2)} = 0 \tag{7.6}$$

these two integrals cancel. The singularity at $\tau = 0$ is by-passed on a quarter circle. The integral along that one is a quarter of the residuum, i.e. $i\pi/2$, in accordance with formula (7.5) for $\sigma = 0$.

The integral I_3 is obtained immediately from I_1 by setting $\sigma = -\zeta$, where $-\zeta$ is again below the integration path and consequently outside of the contour sketched in Fig. 86:

$$I_3 = \int\limits_{\pi/2}^{i\infty}\frac{d\tau}{\sin\tau + \sin\zeta} = i\cdot\frac{\pi/2 - \zeta}{\cos\zeta}. \tag{7.7}$$

For the calculation of the integral I_2, we state that the point $\sigma = \zeta$ is situated *above* the original integration path (compare with Fig. 85) and consequently the residuum has to be taken into account, i.e. a modified relation (7.3)

$$\int\limits_{i\infty}^{\pi/2}\cdots + \int\limits_{\pi/2}^{i\infty+\pi/2}\cdots + \int\limits_{R\infty}\cdots$$

$$= 2\pi i\,\mathrm{Res}\left.\frac{1}{\sin\tau - \sin\zeta}\right|_{\tau=\zeta} = \frac{2\pi i}{\cos\zeta} \tag{7.8}$$

holds and considering (7.5) it follows that

$$\int_{\pi/2}^{i\infty} \frac{d\tau}{\sin\tau - \sin\zeta} = -\frac{2\pi i}{\cos\zeta} + i\,\frac{\pi/2 + \zeta}{\cos\zeta} = -i\cdot\frac{\pi + (\pi/2 - \zeta)}{\cos\zeta}. \tag{7.9}$$

7.2. Perturbation method for complex roots

In the following, Marcuse's method [see MARCUSE (1974)] is placed on a more rigorous foundation. Let $f_1(x)$ and $f_2(x, \eta)$ be two functions with the Taylor expansions

$$\begin{aligned} f_1(x) &= f_1(x_0) + \frac{x - x_0}{1!} f_1'(x_0) + \frac{(x - x_0)^2}{2!} f_1''(x_0) + \cdots, \\ f_2(x, \eta) &= f_2(x_0, 0) + \frac{\partial f_2}{\partial x}(x_0, 0)\cdot(x - x_0) \\ &\quad + \frac{\partial f_2}{\partial \eta}(x_0, 0)\cdot\eta + \frac{1}{2}\frac{\partial^2 f_2}{\partial x^2}(x_0, 0)\cdot(x - x_0)^2 \\ &\quad + \frac{\partial^2 f_2}{\partial x\,\partial\eta}(x_0, 0)\cdot(x - x_0)\,\eta + \frac{1}{2}\frac{\partial^2 f_2}{\partial\eta^2}(x_0, 0)\,\eta^2 + \cdots \end{aligned} \tag{7.10}$$

Let the difference of these functions be $F(x, \eta) = f_1(x) - f_2(x, \eta)$, with the additional requirements that

$$F(x_0, 0) = 0, \qquad \frac{\partial F}{\partial x}(x_0, 0) = 0. \tag{7.11}$$

For small η, $\eta > 0$, we search for complex solutions of $F(x, \eta) = 0$. We apply a perturbation method assuming that x is only slightly different from x_0 and using

$$x = x_0 + (A_1 + iB_1)\,\eta + (A_2 + iB_2)\,\eta^2 + \cdots. \tag{7.12}$$

By inserting (7.12) into the equation $F(x, \eta) = 0$ we obtain

$$F = F_1\eta + F_2\eta^2 + \cdots = 0. \tag{7.13}$$

The equation $F_1 = 0$ yields the condition

$$\frac{\partial f_2}{\partial\eta}(x_0, 0) = 0, \tag{7.14}$$

which is a necessary requirement for the applicability of our method. From $F_2 = 0$ the approximate coefficients of first order are obtained as

$$A_1 = 0, \qquad B_1 = \left[\frac{\dfrac{\partial^2 f_2}{\partial \eta^2}(x_0, 0)}{\dfrac{\partial^2 f_2}{\partial x^2}(x_0, 0) - f_1''(x_0)} \right]^{1/2}. \tag{7.15}$$

The approximate solution $x \approx x_0 + iB_1\eta$ satisfies the equation (7.13) up to terms of the second order inclusively.

7.3. The boundary conditions of the type "loosely-bonded interface"

In section 4.5.4., we have pointed out that the existing discrepancy between experimental and theoretical reflection coefficients for seismic surface waves may be a consequence of the oversimplified model of a vertical discontinuity with boundary conditions of the type "welded contact". Indeed, there is a lot of experimental material referring to the fact that usually a discontinuity is represented by a comparatively narrow zone of dismembered and cracked rocks. In these zones, the velocities of elastic waves are much less than right and left from the discontinuity. Just in the same way as a thin layer over a half-space is simulated by the impedance-like boundary condition (2.2), we can imagine that a more complicated vertical fault zone can be approximated by using convenient new boundary conditions. Pod"yapol'sky (1963) has proved that a "loosely-bonded interface" is a suitable approximation for the configuration described above. The philosophy in it is as follows. The first assumption is that the loosely-bonded interface between two solid half-spaces assures continuity of traction across the interface. The second assumption is that there exists a complex linear relation between slip and local shear stress. The detailed form of the proportionality factor m depends on the model we adopt for the interface. Let the interface be the plane $x_1 = 0$. Then, according to Pod"yapol'sky (1963), the simplest version of these boundary conditions looks like

$$\begin{aligned} &\sigma_{11}^{(1)} = \sigma_{11}^{(2)}, \qquad \sigma_{13}^{(1)} = \sigma_{13}^{(2)}; \qquad u_1^{(1)} = u_1^{(2)}, \\ &u_3^{(1)} + m\sigma_{13}^{(1)} = u_3^{(2)}, \qquad x_1 = 0, \quad \text{for} \quad \text{SV-like motion} \end{aligned} \tag{7.16}$$

and

$$\begin{aligned} &\sigma_{12}^{(1)} = \sigma_{12}^{(2)}; \qquad u_2^{(1)} + m\sigma_{12}^{(1)} = u_2^{(2)}, \\ &x_1 = 0, \quad \text{for} \quad \text{SH-like motion.} \end{aligned} \tag{7.17}$$

Note that these boundary conditions could formally be introduced using a free parameter m (cf. the impedance parameter ε!) as we have done it in the first instance for the impedance-like boundary condition. However, POD"YAPOL'SKY (1963) has shown that (7.16) and (7.17) with

$$m = \lim_{\substack{h \to 0 \\ \mu \to 0}} \frac{h}{\mu} \tag{7.18}$$

can be obtained from the boundary conditions of welded contact by carrying out a limiting process of special kind, where h is the thickness of a vertical layer between the two half-spaces (or quarter-spaces in our case) (1) and (2), and μ is its rigidity. The limiting process has to be carried out in such a way that m is always finite. For $m \to 0$, we obtain a welded interface, and infinitely large values of m correspond to an ideally smooth interface. It should be noted that the energy conservation law does not hold any longer for boundary conditions of this type.

Quite recently, following an idea of Yanovskaya, these boundary conditions were for the first time applied to seismic surface wave propagation by ITS and MALISCHEWSKY (1986a, b). The encouraging results show that, depending on m, the reflection coefficients are much higher than for the welded contact. Of course, the difficulty lies in adjusting the parameter m reasonably to the conditions in the Earth. Nevertheless, using these or similar boundary conditions, new interesting results can be expected in the future.

8. References

ABRAM, R. A.: A coupled-mode formalism for elastic waveguides. J. Phys., D. 7 (1974), 1329–1335.

ACHENBACH, J. D.; GAUTESEN, A. K., MENDELSOHN, D. A.: IEEE Trans. Sonics and Ultrason. SU-27 (1980), 124–129.

AKI, K., RICHARDS, P. G.: Quantitative seismology. W. H. Freeman and Company, San Francisco 1980.

ALSOP, L. E.: Transmission and reflection of Love waves at a vertical discontinuity. J. geophys. Res. 71 (1966), 3969–3984.

ALSOP, L. E., GOODMAN, A. S., ASH, E.: Surface wave dispersion in a mass-loaded half-space. J. Acoust. Soc. Am. 50 (1971), 176–180.

ALSOP, L. E., GOODMAN, A. S., GREGERSEN, S.: Reflection and transmission of inhomogeneous waves with particular application to Rayleigh waves. Bull. Seism. Soc. Am. 64 (1974), 1635–1652.

BABICH, V. M., MOLOTKOV, I. A.: Propagation of Love waves in an elastic half-space which is inhomogeneous in two coordinates (In Russian). Izv. Akad. Nauk SSSR, Fiz. Zemli (1966) 6, 34–38.

BOORE, D. M.: Love waves in non-uniform wave guides: finite difference calculations. J. geophys. Res. 75 (1970), 1512–1527.

BOORE, D. M., HARMSEN, S. C., HARDING, S. T.: Wave scattering from a step change in surface topography. Bull. Seism. Soc. Am. 71 (1981), 117–125.

BREKHOVSKIKH, L. M.: Waves in layered media. Academic Publishers, New York 1961.

BROOKE, G. H., KHARADLY, M. M. Z.: Step discontinuities on dielectric waveguides. Electron. Lett. 12 (1976), 473–475.

BUDDEN, K. G.: The wave-guide mode theory of wave propagation. Logos Press, London 1961.

BUKCHIN, B. G., LEVSHIN, A. L.: Propagation of Love waves across a vertical discontinuity. Wave Motion 2 (1980), 293–302.

CARON, A.: Numerical methods of solution of linear and nonlinear integral equations (In French). Ph. D. thesis, L'Universite Pierre et Marie Curie, Paris 1981.

CHAMUEL, J. R.: New ultrasonic modeling findings on Rayleigh wave propagation and their implications. J. Acoust. Soc. Am. 75 (1984), 1495–1504.

CHANG, C. P., TUAN, H. S.: On the surface-to-bulk mode conversion of Rayleigh waves. IEEE Trans. Microwave Theory Tech. MTT-21 (1973), 558–560.

CHEN, T. C., ALSOP, L. E.: Reflection and transmission of obliquely incident Rayleigh waves at a vertical discontinuity between two welded quarter-spaces. Bull. Seism. Soc. Am. **69** (1979), 1409–1423.

Clayton, E., Derrick, G. H.: A numerical solution of wave equations for real and complex eigenvalues. Austr. J. Phys. **30** (1977), 15–21.

Collin, R. E.: Field theory of guided waves. Mc Graw-Hill, New York 1960.

Davies, J. B.: A least-squares boundary residual method for the numerical solution of scattering problems. IEEE Trans. Microwave Theory Tech. MTT-21 (1973), 99–104.

Dow, M. L.: Singular integral equations. Ph. D. thesis, University of Tasmania, Hobart, Australia, 1977.

Dow, M. L., Elliott, D.: The numerical solution of singular integral equations over (−1, 1). SIAM J. Numer. Anal. **16** (1979), 115–134.

Drake, L. A.: Love and Rayleigh waves in an irregular soil layer. Bull. Seism. Soc. Am. **70** (1980), 571–582.

Drake, L. A., Bolt, B. A.: Love waves normally incident at a continental boundary. Bull. Seism. Soc. Am. **70** (1980), 1103–1123.

Dresen, L., Freystätter, S.: Rayleigh channel waves for the in-seam seismic detection of discontinuities. J. Geophys. **42** (1976), 111–129.

England, R., Sabina, F. J., Herrera, I.: Scattering of SH waves by surface cavities of arbitrary shape using boundary methods. Phys. Earth planet. Interiors **21** (1980), 148–157.

Fain, W. M., Chanin, J. I.: Quantum electronics (In German). B. G. Teubner Verlagsgesellschaft, Leipzig 1969.

Fedoryuk, M. V.: Foundation of the cross-sectional method for inhomogeneous acoustic wave-guides (In Russian). Zh. vychisl. Mat. mat. Fiz. **13** (1973), 127–135.

Felsen, L. B., Marcuvitz, N.: Radiation and scattering of waves. Prentice-Hall, Englewood Cliffs, New Yersey, 1973.

Freund, L. B.: Guided surface waves on an elastic half space. J. Appl. Mech. **38** (1971), 899–905.

Fuyuki, M.; Matsumoto, Y.: Finite difference analysis of Rayleigh wave scattering at a trench. Bull. Seism. Soc. Am. **70** (1980), 2051–2069.

Gegechkori, T. Sh., Yanovskaya, T. B.: Reflection of Rayleigh waves at the border of contact of media with strong and weak velocity contrast (In Russian). Soobshch. Akad. Nauk Gruz. SSR **96** (1979), 573–576.

Gelin, P., Petenzi, M., Citerne, J.: Rigorous analysis of the scattering of surface waves in an abruptly ended slab dielectric waveguide. IEEE Trans. Microwave Theory Tech. MTT-29 (1981), 107–114.

Goncharenko, A. M., Karpenko, V. A.: Fundamentals of the theory of optical waveguides (In Russian). Nauka i Tekhnika, Minsk 1983.

Gregersen, S.: Amplitudes of horizontally refracted Love waves. Ph. D. Thesis, Columbia University, New York 1974.

Gregersen, S.: Possible mode conversion between Love and Rayleigh waves at a continental margin. Geophys. J. R. astr. Soc. **54** (1978), 121–127.

Gregersen, S.: L_g-wave propagation and structure differences near Denmark and the North Sea. Geophys. J. R. astr. Soc. **79** (1984), 217–234.

Gregersen, S., Alsop, L. E.: Amplitudes of horizontally refracted Love waves. Bull. Seism. Soc. Am. **64** (1974), 535–553.

Gregersen, S., Alsop, L. E.: Mode conversion of Love waves at a continental margin. Bull. Seism. Soc. Am. **66** (1976), 1855–1872.

Haus, H. A., Lattes, A., Melngailis, J.: Grating coupling between surface acoustic waves and plate modes. IEEE Trans. Sonics and Ultrasonics SU-**27** (1980), 258 to 265.

Herrera, I.: On a method to obtain a Green's function for a multi-layered half-space. Bull. Seism. Soc. Am. **54** (1964), 1087–1096.

Herrera, I.: Theory of connectivity for formally symmetric operators. Proc. Nat. Acad. Sci. Am. **74** (1977), 4722–4725.

Herrera, I.: Theory of connectivity: a systematic formulation of boundary element methods. Appl. Math. Modelling **3** (1979), 151–156.

Herrera, I.: An algebraic theory of boundary value problems. Kinam (Testimonia de Fisica), **3** (1981), 161–229.

Herrera, I.: Instituto de Geofisica, Mexico, Personal communication 1982.

Herrera, I., Spence, D. A.: Framework of biorthogonal Fourier series. Proc. Nat. Acad. Sci. Am. 78 (1981), 7240–7244.

Higuchi, S.: On Love dispersion in a complicated surface layer (In Japanese). Zisin. **4** (1932), 271–276.

Hudson, J. A.: The passage of elastic waves through an anomalous region — IV. Transmission of Love waves through a laterally varying structure. Geophys. J. R. astr. Soc. **49** (1977), 645–654.

Ioakimidis, N. I.: An iterative algorithm for the numerical solution of singular integral equations. J. comput. Phys. **43** (1981), 164–176.

Irikura, K., Kawanaka, T.: Characteristics of microtremors on ground with discontinuous underground structure. Bull. Disast. Prev. Res. Inst., Kyoto Univ., **30** (1980), 81–96.

Iskander, K. A., Hamid, M. A. K.: Ray optical analysis of waveguide transformers with dielectric inserts. Int. J. Electronics **39** (1975), 481–490.

Its, E. N., Yanovskaya, T. B.: Reflection and refraction of Rayleigh waves at a vertical discontinuity (In Russian). Vychisl. Seismol. **10** (1977), 214–222.

Its, E. N., Yanovskaya, T. B.: Reflection and refraction of surface waves for oblique incidence at a vertical discontinuity (In Russian). Vychisl. Seismol. **12** (1979a), 86–92.

Its, E. N., Yanovskaya, T. B.: Determination of reflection and transmission coefficients of surface waves at a vertical contact using Green's function (In Russian). Izv. Akad. Nauk SSSR, Fiz. Zemli (1979b) 6, 11–21.

Its, E. N., Yanovskaya, T. B.: The use of the refined Green's function in evaluating reflection and transmission coefficients of Rayleigh waves on a vertical boundary (In Russian). Vychisl. Seismol. **14** (1982), 147–152.

Its, E. N., Yanovskaya, T. B.: Propagation of surface waves in a half-space with vertical, inclined or curved interfaces. Wave Motion **7** (1985), 79–94.

Ittipiboon, A., Hamid, M.: Scattering of surface waves at a slab waveguide discontinuity. Proc. IEE **126** (1979), 798–804.

ITTIPIBOON, A., HAMID, M.: Application of the Wiener-Hopf-technique to dielectric slab waveguide discontinuities. IEE Proc., Pt. H 128 (1981), 188—196.

JÖRGENS, K., RELLICH, F.: Theory of eigenvalues of ordinary differential equations (In German). Springer-Verlag, Berlin 1976.

KAMEL, A., FELSEN, L. B.: Hybrid ray-mode formulation of SH-motion in a two-layer half-space. Bull. Seism. Soc. Am. 71 (1981), 1763—1781.

KATSENELENBAUM, B. Z.: Theory of irregular wave-guides with slowly varying parameters (In Russian). Izdatel'stvo Akademii Nauk SSSR, Moskva 1961.

KAY, A. F.: Scattering of a surface wave by a discontinuity in reactance. IRE Trans. on Antennas and Prop. AP-7 (1959), 22—31.

KAZI, M. H.: Spectral representation of the Love wave operator. Geophys. J. R. astr. Soc. 47 (1976), 225—249.

KAZI, M. H.: The Love wave scattering matrix for a continental margin (theoretical). Geophys. J. R. astr. Soc. 52 (1978a), 25—44.

KAZI, M. H.: The Love wave scattering matrix for a continental margin (numerical). Geophys. J. R. astr. Soc. 58 (1978b), 227—243.

KAZI, M. H.: Reciprocity relations between transmission coefficients in Love wave propagation. Dep. of Math. Sci., University of Petroleum and Minerals, Dhahran, Saudi-Arabia, Technical Report No. 6, 1980.

KAZI, M. H., NIAZY, A.: Diffraction of Love waves at a vertical step up and step down discontinuity I. Dep. of Math. Sci., University of Petroleum and Minerals, Dhahran, Saudi-Arabia, Technical Report No. 38, 1982a.

KAZI, M. H., NIAZY, A.: Generalization of Higuchi's conditions for Love waves propagating through two welded quarterspaces with two surface layers on each. Dep. of Math. Sci., University of Petroleum and Minerals, Dhahran, Saudi-Arabia, Technical Report No. 45, 1982b.

KAZI, M. H., NIAZY, A.: The Love wave scattering matrix for three-layered structures consisting of welded layered quarterspaces with a plane surface. Dep. of Math. Sci., University of Petroleum and Minerals, Dhahran, Saudi-Arabia, Technical Report No. 70, 1984.

KAZI, M. H., WILLMER, D. G.: On the spectral representation of the Rayleigh wave operator — I. Mathematical formulation. Dep. of Math. Sci., University of Petroleum and Minerals, Dhahran, Saudi-Arabia, Technical Report No. 29, 1981.

KENNETT, B. L. N.: Guided wave propagation in laterally varying media — I. Theoretical development. Geophys. J. R. astr. Soc. 79 (1984), 235—255.

KENNETT, B. L. N., MYKKELTVEIT, S.: Guided wave propagation in laterally varying media — II. L_g-waves in North-Western Europe. Geophys. J. R. astr. Soc. 79 (1984), 257—267.

KNOPOFF, L., MAL, A. K.: Phase velocity of surface waves in the transition zone of continental margins. J. geophys. Res. 72 (1967), 1769—1776.

KNOPOFF, L., MAL, A. K., ALSOP, L. E., PHINNEY, R. A.: A property of long-period Love waves. J. geophys. Res. 75 (1970), 4084—4086.

KORN, M., STÖCKL, H.: Reflection and transmission of Love channel waves at coal seam discontinuities computed with a finite difference method. J. Geophys. 50 (1982), 171—176.

KOSHIBA, M., MORITA, H., SUZUKI, M.: Finite-element analysis of discontinuity problem of SH modes in an elastic plate waveguide. Electron. Lett. **17** (1981), 480–482.

KUDO, M., MUSHIAKE, Y.: Mode selecting characteristics of four-layer dielectric slab waveguide. Radio Sci. **17** (1982), 125–134.

LANDAU, L. D., LIFSCHITZ, E. M.: Course of theoretical physics, volume III, Quantum mechanics (In German). Akademie-Verlag, Berlin 1965.

LEVSHIN, A. L.: Effect of lateral inhomogeneity on surface wave measurements (In Russian). Vychisl. Seismol. **16** (1984), 118–127.

LEVSHIN, A. L., YANOVSKAYA, T. B.: Reflection and refraction of Love waves at a vertical discontinuity (In Russian). Vychisl. Seismol. **9** (1976), 160–173.

LI, R. C. M., MELNGAILIS, J.: The influence of stored energy at step discontinuities on the behaviour of surface-wave gratings. IEEE Trans. Sonics and Ultrason. SU-**22** (1975), 189–198.

LOVE, A. E. H.: Some problems of geodynamics. Cambridge University Press, 1911.

LUTIKOV, A. I.: Influence of an inclined discontinuity on the reflection and refraction of Rayleigh waves (In Russian). Izv. Akad. Nauk SSSR, Fiz. Zemli (1979) **10**, 58–66.

LYSMER, J., DRAKE, L. A.: The propagation of Love waves across non-horizontally layered structures. Bull. Seism. Soc. Am. **61** (1971), 1233–1251.

MAHMOUD, S. F., BEAL, J. C.: Scattering of surface waves at a dielectric discontinuity on a planar waveguide. IEEE Trans. Microwave Theory Tech. MTT-**23** (1975), 193–198.

MALISCHEWSKY, P.: Orthonormalization of plane surface and body waves (In German). Gerl. Beitr. Geophys. **79** (1970), 468–474.

MALISCHEWSKY, P.: Consideration of certain singularities in Haskell's matrix method. Gerl. Beitr. Geophys. **80** (1971), 457–462.

MALISCHEWSKY, P.: Propagation of seismic surface waves in media with vertical discontinuities (In German). Veröff. Zentlnst. Phys. Erde, No. 24, AdW der DDR, Potsdam 1973.

MALISCHEWSKY, P.: The influence of curved discontinuities on the propagation of seismic surface waves. Gerl. Beitr. Geophys. **83** (1974), 355–362.

MALISCHEWSKY, P.: Surface waves in media having lateral inhomogeneities. Pure Appl. Geophys. **114** (1976), 833–843.

MALISCHEWSKY, P.: The influence of vertical discontinuities on the propagation of seismic surface waves, in: Theory of diffraction and propagation of waves, VII. All-union symposium about diffraction and propagation of waves. Rostov on the Don, USSR, Abstracts of the papers, 386–390, Moscow 1977.

MALISCHEWSKY, P.: An improvement of Alsop's method for the determination of reflection and transmission coefficients of surface waves (In German). Pure Appl. Geophys. **117** (1979), 1045–1049.

MALISCHEWSKY, P.: Application of the improved version of Alsop's method to the passage of Rayleigh waves through a vertical discontinuity (In Russian). Izv. Akad. Nauk SSSR, Fiz. Zemli (1980) **11**, 87–89.

MALISCHEWSKY, P.: A semi-analytical method for the calculation of leaking Love-wave modes. Wave Motion **7** (1985), 253–262.

Malischewsky, P., Neunhöfer, H.: Observations of surface waves in the southern part of GDR and possibilities for a theoretical interpretation (In German). Freiberger Forschhft. C 387 (1983), 51–60.

Manenkov, A. B.: Propagation of a surface wave through an inhomogeneous part of a thin dielectric wave-guide (In Russian). Radioteknika i elektronika **23** (1978), 33–40.

Maradudin, A. A.: Surface waves, in: Festkörperprobleme (Advances in Solid State Physics). Vol. XXI, 25–116, Vieweg, Braunschweig 1981.

Marcuse, D.: Radiation losses of tapered dielectric slab waveguides. Bell Syst. Techn. J. **49** (1970), 273–290.

Marcuse, D.: Light transmission optics. Van Nostrand Reinhold Company, New York 1972.

Marcuse, D.: Theory of dielectric optical waveguides. Academic Press, New York 1974.

Martel, L.: Love wave propagation across a step by finite elements and spatial filtering. Geophys. J. R. astr. Soc. **61** (1980), 659–678.

Matthews, H. (Ed.): Surface wave filters. John Wiley & Sons, New York 1977.

McGarr, A.: Amplitude variations of Rayleigh waves — propagation across a continental margin. Bull. Seism. Soc. Am. **59** (1969a), 1281–1305.

McGarr, A.: Amplitude variations of Rayleigh waves — horizontal refraction. Bull. Seism. Soc. Am. **59** (1969b), 1307–1334.

McGarr, A.; Alsop, L. E.: Transmission and reflection of Rayleigh waves at vertical boundaries. J. geophys. Res. **72** (1967), 2169–2179.

Mendelsohn, D. A., Achenbach, J. D., Keer, L. M.: Scattering of elastic waves by a surface-breaking crack. Wave Motion **2** (1980), 277–292.

Millar, R. F.: The Rayleigh hypothesis and a related least-squares solution to scattering problems for periodic surfaces and other scatterers. Radio Sci. **8** (1973), 785 to 796.

Miller, M. A., Talanov, V. I.: Application of the surface-impedance concept in the theory of electromagnetic surface waves (In Russian). Izv. vuzov, Radiofizika **4** (1961), 795–830.

Mitchell, B. J., Leite, L. W. B., Yu, Y. K., Herrmann, R. B.: Attenuation of Love and Rayleigh waves across the Pacific at periods between 15 and 110 seconds. Bull. Seism. Soc. Am. **66** (1976), 1189–1201.

Mittra, R., Lee, S. W.: Analytical techniques in the theory of guided waves. The Mac Millan Company, New York 1971.

Molotkov, L. A.: Investigation of interferring waves in layered media (In Russian). Ph. D. Thesis, University of Leningrad, 1974.

Morishita, K., Inagaki, S., Kumagai, N.: Analysis of discontinuities in dielectric waveguides by means of the least squares boundary residual method. IEEE Trans. Microwave Theory Tech. MTT-27 (1979), 310–315.

Munasinghe, M., Farnell, G. W.: Finite difference analysis of Rayleigh wave scattering at vertical discontinuities. J. geophys. Res. **78** (1973), 2454–2466.

Murphy, J. R., O'Brien, L. J.: An investigation of the reflection of short-period surface waves from valley boundaries, represented at the Annual Meeting of the Seismological Society of America, 1978.

MUSKHELISHVILI, N. I.: Singular integral equations. P. Noordhoff N. V., Groningen 1953.

NEFEDOV, E. I.: Diffraction of electromagnetic waves at dielectric structures (In Russian). Nauka, Moskva 1979.

NEUMARK, M. A.: Linear differential operators (In German). Akademie-Verlag, Berlin 1960.

NEUNHÖFER, H.: Primary and secondary effects of surface wave propagation and their geophysical causes (In German). Veröff. Zentlnst. Phys. Erde No. 85, AdW der DDR, Potsdam 1985.

NEUNHÖFER, H., GÜTH, D., MALISCHEWSKY, P.: Investigation of seismic surface waves with regard to connections with geodynamics (In German). Veröff. Zentlnst. Phys. Erde No. 29, AdW der DDR, Potsdam 1974.

NIAZY, A., KAZI, M. H.: On the Love wave scattering problem for welded layered quarter-spaces with applications. Bull. Seism. Soc. Am. **70** (1980), 2071—2095.

NIAZY, A., KAZI, M. H.: On the effect of the higher modes on the scattering of Love waves at the boundary of welded layered quarter-spaces. Dep. of Math. Sci., University of Petroleum and Minerals, Dhahran, Saudi-Arabia, Technical Report No. 27, 1981.

OGUSU, K., MIYAGI, M., NISHIDA, S.: Leaky TE modes on an asymmetric three-layered slab waveguide. J. Opt. Soc. Am. **70** (1980), 48—52.

PITTS, L. E., PLONA, T. J., MAYER, W. G.: Theoretical similarities of Rayleigh and Lamb modes of vibration. J. Acoust. Soc. Am. **60** (1976), 374—377.

PUDENSI, M. A. A., FERREIRA, L. G.: Method to calculate the reflection and transmission of guided waves. J. Opt. Soc. Am. **72** (1982), 126—130.

RAYLEIGH, Lord: On waves propagated along the plane surface of an elastic solid. Proc. Lond. Math. Soc. **17** (1887), 4—11.

ROKHLIN, S.: Diffraction of Lamb waves by a finite crack in an elastic layer. J. Acoust. Soc. Am. **67** (1980), 1157—1165.

ROZZI, T. E.: Rigorous analysis of the step discontinuity in a planar dielectric waveguide. IEEE Trans. Microwave Theory Tech. MTT-**26** (1978), 738—746.

RULF, B.: Surface-wave diffraction and guidance. J. Acoust. Soc. Am. **62** (1977a), 1110—1115.

RULF, B.: On scattering of elastic surface waves by edges of thin surface layers at oblique incidence. J. Acoust. Soc. Am. **62** (1977b), 1300—1302.

RULF, B.: On the matching of two optical waveguides. Radio Sci. **12** (1977c), 593—601.

RULF, B., KEDEM, N.: Surface wave diffraction be means of singular integral equations. J. Sound Vibr. **45** (1976), 15—28.

RYKUNOV, L. N., FAM VAN TKHUKH: Investigation of the influence of local inhomogeneities on the Rayleigh-wave field (In Russian). Izv. Akad. Nauk SSSR, Fiz. Zemli (1972) 5, 65—71.

SAASTAMOINEN, P. R.: Ray asymptotic propagation of the spectra of elastic surface waves in a waveguide with laterally smooth inhomogeneities. Geophys. J. R. astr. Soc. **79** (1984), 269—284.

SATÔ, R.: Love waves propagated across transitional zone. Jap. J. Geophys. **2** (1961), 117—134.

Schlue, J. W.: Love-wave propagation in three-dimensional structures using finite element techniques. Bull. Seism. Soc. Am. **69** (1979), 319–328.

Schlue, J. W.: Seismic surface wave propagation in three-dimensional finite-element structures. Bull. Seism. Soc. Am. **71** (1981), 1003–1010.

Sezawa, K.: Dispersion of elastic waves propagated on the surface of stratified bodies and on curved surfaces. Bull. Earthq. Res. Inst. Tokyo Univ. **3** (1927), 1–18.

Sezawa, K., Kanai, K.: The M_2 seismic waves. Bull. Earthq. Res. Inst. Tokyo Univ. **13** (1935), 471–475.

Shen, T. P., Maradudin, A. A., Wallis, R. F., Stegeman, G. I.: Transmission and reflection of a surface acoustic wave through a transverse boundary separating two isotropic media. University of California, Irvine, Technical Report No. 84-16, 1984.

Shevchenko, V. V.: Continuous transitions in open wave-guides. Golem Press, Boulder, Colorado, 1971a.

Shevchenko, V. V.: Decomposition of fields in open wave-guides in terms of proper and improper eigenfunctions (In Russian). Izv. vuzov, Radiofizika **14** (1971b), 1242 to 1249.

Shevchenko, V. V.: Institute for radio engineering and electronics of the Academy of Sciences of USSR. Moscow, Personal communication 1978.

Shevchenko, V. V.: Spectral decomposition in terms of associated eigenfunctions of a non-self-adjoint Sturm-Liouville problem on the whole axis (In Russian). Diff. urav. **15** (1979), 2004–2020.

Sikharulidze, D. I.: A study of reflected surface waves. The Ninth Assembly of E. S. C., Copenhagen 1967.

Sikharulidze, D. I.: Composition of the Earth from surface wave data (In Russian). Metsniereba, Tbilissi 1978.

Simons, D. A.: Scattering of a Love wave by the edge of a thin surface layer. J. Appl. Mech. **42** (1975), 842–846.

Simons, D. A.: Scattering of a Rayleigh wave by the edge of a thin surface layer of negligible inertia. J. Acoust. Soc. Am. **59** (1976), 12–18.

Smirnov, V. I.: Course of higher mathematics part IV (In German). Deutscher Verlag der Wissenschaften, Berlin 1975.

Stegeman, G. I., Maradudin, A. A., Rahman, T. S.: Refraction of a surface polariton by an interface. Phys. Rev. B **23** (1981), 2576–2585.

Stoneley, R.: Elastic waves at the surface of separation of two solids. Proc. R. Soc., A. **106** (1924), 416–428.

Talanov, V. I.: Diffraction of electromagnetic waves at a surface-impedance jump in a wave-guide (In Russian). Izv. vuzov. Radiofizika **1** (1958), 64–72.

Tien, P. K., Smolinsky, G., Martin, R. J.: Radiation fields of a tapered film and a novel film-to-fiber coupler. IEEE Trans. Microwave Theory Tech. MTT-**23** (1975), 79–85.

Tiersten, H. F.: Elastic surface waves guided by thin films. J. Appl. Phys. **40** (1969), 770–789.

Toksöz, M. N., Bullitt, J. T.: Three-dimensional ultrasonic modeling of Rayleigh wave scattering. EOS. Trans. Am. geophys. Un. **65** (1984), 234.

TORVIK, P. J.: Reflection of wave trains in semi-infinite plates. J. Acoust. Soc. Am. 41 (1967), 346—353.

TUAN, H.-S., CHANG, CH.-P.: Tapping of Love waves in an isotropic surface waveguide by surface-to-bulk wave transduction. IEEE Trans. Microwave Theory Tech. MTT-20 (1972), 472—477.

TUAN, H. S., LI, R. C. M.: Rayleigh-wave reflection from groove and step discontinuities. J. Acoust. Soc. Am. 55 (1974), 1212—1217.

UZUNOGLU, N. K.: Scattering from inhomogeneities inside a fiber waveguide. J. Opt. Soc. Am. 71 (1981), 259—273.

WATSON, T. H.: High and low frequency behaviour of leaking modes in the layered half-space. Ph. D. Thesis, University of Pittsburg, 1970.

WEIDMANN, J.: Spectral theory of Sturm-Liouville operators (In German). Math. Z. 98 (1967), 268—302.

WEIDMANN, J.: Linear operators in Hilbert space (In German). Teubner-Verlag, Stuttgart 1976.

WEIDMANN, J.: Johann-Wolfgang-Goethe-Universität, Frankfurt am Main, F. R. G., Fachbereich Mathematik, Personal communication 1982.

WEINSTEIN, L. A.: The theory of diffraction and the factorization method. Golem Press, Boulder/Colorado 1969.

WIENER, N., HOPF, E.: About a class of singular integral equations (In German). Preuss. Akad. Wiss., Phys. Math. Kl. (1931), 696—706.

WILCOX, C. H.: Scattering theory for the d'Alembert equation in exterior domains. Springer-Verlag, Berlin 1975.

WILLMER, D. G., KAZI, M. H.: On the generalized eigenvalue problem of free harmonic motion in infinite elastic layers. Dep. of Math. Sci., University of Petroleum and Minerals, Dhahran, Saudi-Arabia, Technical Report No. 61, 1983.

WOODHOUSE, J. H.: Surface waves in a laterally varying structure. Geophys. J. R. astr. Soc. 37 (1974), 461—490.

WORKSHOP: Report on the workshop on application of elastic waves in electrical devices, non-destructive testing and seismology. Northwestern University, Evanston, Illinois, U.S.A., 1976.

YASUURA, K., SHIMOHARA, K., MIYAMOTO, T.: Numerical Analysis of a thin-film waveguide by mode matching method. J. Opt. Soc. Am. 70 (1980), 183—191.

YONEYAMA, T., NISHIDA, S.: Reflection and transmission of Rayleigh waves by the edge of a deposited thin film. J. Acoust. Soc. Am. 55 (1974), 738—743.

YONEYAMA, T., NISHIDA, S.: Approximate solution for step discontinuity in dielectric slab waveguide. Electron. Lett. 17 (1981), 151—153.

ZAVADSKY, V. YU.: Calculation of wave fields in open regions and wave-guides (In Russian). Nauka, Moskva 1972.

Additional References

BULLITT, J. T., TOKSÖZ, M. N.: Three-dimensional ultrasonic modeling of Rayleigh wave propagation. Bull. Seism. Soc. Am. 75 (1985), 1087—1104.

CARON, A., DUPUY, M., PICHOT, CH.: Numerical solution of singular integral equations.

Application to the flanged plane waveguide. J. comput. Phys. **38** (1980), 357 to 377.

CHANG, D. C., KUESTER, E. F.: Total and partial reflection from the end of a parallel-plate waveguide with an extended dielectric slab. Radio Sci. **16** (1981), 1–13.

COHEN, H.: Accurate numerical solutions of integral equations with kernels containing poles. J. comput. Phys. **26** (1978), 257–276.

COURANT, R., HILBERT, D.: Methods of mathematical physics I (In German). Springer-Verlag, Berlin 1968, 360–363.

EZAWA, H., KAWAJI, S., NAKAMURA, K.: Surfons and the electron mobility in silicon inversion layers. Japan. J. Appl. Phys. **13** (1974), 126–155.

ITS, E. N., MALISCHEWSKY, P.: Propagation of Rayleigh waves through a loosely-bonded vertical interface of elastic media (In Russian). Izv. Akad. Nauk SSSR, Fiz. Zemli (1986a), submitted for publication.

ITS, E. N., MALISCHEWSKY, P.: Reflection and transmission of Love waves across a vertical boundary of horizontally homogeneous media in the case of loosely-bonded contact (In Russian). Gerl. Beitr. Geophys. (1986b), submitted for publication.

KAZI, M. H., ABU-SAFIYA, A. S. M.: Spectral representation of the Love wave operator for two layers over a half-space. Dep. of Math. Sci., University of Petroleum and Minerals, Dhahran, Saudi-Arabia, Technical Report No. 30, 1981.

LEVSHIN, A. L., YANOVSKAYA, T. B., LANDER, A. V., BUKCHIN, B. G., BARMIN, M. P., RATNIKOVA, L. L., ITS, E. N.: Seismic surface waves in the horizontally inhomogeneous Earth (In Russian). Nauka, Moskva 1986.

MOMOI, T., Scattering of Rayleigh wave of a vertical boundary. J. Phys. Earth **29** (1981), 435–485.

POD"YAPOL'SKY, G. S.: Reflection and refraction at a boundary of two elastic media in the case of loosely-bonded contact (In Russian). Izv. Akad. Nauk SSSR, Ser. Geofiz. (1963) **4**, 525–531.

RADOVICH, B. J., DE BREMAECKER, J. C.: Body waves as normal and leaking modes — leaking modes of Love waves. Bull. Seism. Soc. Am. **64** (1974), 301–306.

VIKTOROV, I. A.: Acoustic surface waves in solids (In Russian). Nauka, Moskva 1981, 88–94.

Subject Index

Author Index